Concepts in Catalysis

1918: *Πάντα ῥεῖ κουδὲν ἐν ταὐτω μένει*

1968: *Eppur si muove*

Concepts in Catalysis

Eric K. Rideal
Formerly Professor of Colloid Science
University of Cambridge, England

1968
Academic Press London & New York

ACADEMIC PRESS INC. (LONDON) LTD
Berkeley Square House
Berkeley Square
London W.1

U.S. Edition published by

ACADEMIC PRESS INC.
111 Fifth Avenue
New York
New York 10003

Library of Congress Catalog Card Number: 68-17669

Printed in Great Britain by
Spottiswoode, Ballantyne and Company Limited
London and Colchester

Foreword

A person is confronted with many problems on entering that twilight zone when, owing to superannuation from a University, one is of necessity suspended between the actively temporal world and that of the timeless spirit. The mere mechanisms of living are complicated enough in a country famed for its commissions and committees investigating geriatry, now regarded as a "subject" suitable for scientific study; however, these pale into insignificance compared to the problems of the mind. A laboratory in which research is being actively pursued develops a spirit. Indeed Pasteur likened a laboratory to a temple animated by the spirit of enthusiasm ἐν θεος—the god within.

The realization that one must leave Eden and face a change is hard to bear:

> Halte nur Dein Rosz im Zügel
> kommst ja doch nicht allzu weit
> hinter jedem neuen Hügel
> dehnt sich die Unendlichkeit

The blow has been softened for me to a remarkable degree by the kindness of the authorities of Imperial College in receiving me into their midst. I am especially indebted to Professor R. M. Barrer, the Head of the Department of Chemistry, who has done all things possible for me to continue my passage amongst colleagues and friends.

ERIC K. RIDEAL

September 1967
London

Acknowledgments

I have drawn freely both from laboratory work and from the following compilations: (1948–) *Advances in Catalysis*, Academic Press, New York; (1954–) *Catalysis*, Reinhold, New York; 1st (1956), 2nd (1957), 3rd (1964) International Congress on Catalysis; G. C. Bond (1962) *Catalysis by Metals*, Academic Press, London; S. Brunauer (1943) *The Adsorption of Gases and Vapors*, Princeton University Press, New Jersey; D. O. Hayward and D. M. W. Trapnell (1962) *Chemisorption*, Butterworths, London; F. F. Volkenshtein (1963) *The Electronic Theory of Catalysis on Semi-conductors*, Pergamon Press, Oxford.

Contents

Chapter I

Introduction

Fifty years ago Sir Hugh Taylor and I had the opportunity of writing *Catalysis in Theory and Practice*, in which we attempted to give an account of the catalytic phenomena known at the time. During the intervening years both the scientific inquiries as to the "modus operandi" of catalytic action and the technical developments in catalytic industrial processes have grown at an ever accelerating pace. This growth has been greatly stimulated by the utilization of new tools either chemical, such as the ortho–para hydrogen, deuterium and tritium reactions, labelling with radioisotopes and the development of knowledge of ligand structure in organometallic compounds, or physical, involving the methods of detection of free radicals and the properties of semiconductors, as well as the various magnetic, optical and electronic devices for examination of catalytic systems. On the industrial side the processes of ammonia synthesis and oxidation as well as the hydrogenation of unsaturated oils are now so firmly established as fundamental in world economy and the ever expanding range of chemicals provided by the petrochemical industry so important as to ensure that research on catalytic mechanisms will not be confined to academic laboratories.

The growth is revealed throughout the world in industrial plants, in Institutes devoted to the study of catalysis and to a truly frightening expansion of the literature on the subject. The compilation of a book on heterogeneous catalysis would consequently present a very formidable task. It is for personal reasons that I have ventured to put pen to paper in an attempt to portray some aspects of the subject which have impressed me. As an undergraduate and in my early graduate years at Cambridge, I was deeply influenced by Sir William Hardy and later by Irving Langmuir. A belated tribute to the memory of two lifelong friends might not be out of place in acknowledgment of my indebtedness to them. If any more formal dedication is required it is to those working in the laboratory at Cambridge who were sufficiently intrigued with the problems of heterogeneous catalysis to maintain their contributions to this subject when they went out to take their several parts in the life of the scientific community. These include R. M. Barrer, R. C. L. Bosworth, F. P. Bowden, D. D. Eley and A. and L. Farkas, E. F. G. Herington, C. Kemball, H. W. Melville, J. K. Roberts, H. Steiner, B. W. M. Trapnell, G. H. Twigg and A. F. H. Ward. As is clear from the title of this book, I have not attempted to write a comprehensive volume on heterogeneous catalysis but have drawn largely from the work of my colleagues and the material of my lectures to discuss some of the factors which may play a part in these surface chemical reactions. I pray forgiveness for many sins of omission and commission in the text.

In looking over the manuscript I note that several of the views expressed appear to be mutually exclusive. This may be due in part to a deceptive simplicity apparent in many reactions. The introduction of hydrogen isotopes and of the ortho–para reaction into catalytic reactions of the hydrocarbons has thrown a great deal of light on the mechanisms involved but has raised as many questions as it set out to solve. Again, there is a considerable latitude in the interpretation of experimental data in terms of mechanism. Preconceived views or choice of analogies may play a dominant part in one's opinion. As an example we note that data both recent and accurate on the adsorption isotherms of hydrogen on carbon have been interpreted both as an adsorption on fixed sites with lateral interaction and as a two-dimensional Van der Waals gaseous state. The classical example of a heterogeneous catalytic reaction, namely the hydrogenation of ethylene (a system which possibly has been studied more than any other), still provides an area for vigorous debate. In the field of catalysis the crucial experiment is often a will o' the wisp and eludes our grasp.

Fifty years ago, two contrasting views on the regulatory mechanisms were firmly held: that of Sabatier in the concept of the formation of intermediary chemical compounds and that of Faraday, revived by Bodenstein, in the theory of diffusive control. Langmuir's development of the idea of the orientated monolayer and the recognition of the distinction between physically adsorbed and chemisorbed species by H. S. Taylor in his work on activated adsorption greatly supported the chemical view and the conditions necessary for limitation by diffusive control in any given reaction were readily established.

During the intervening years, our knowledge concerning processes of condensation and evaporation has been greatly extended and we have witnessed the development of electronic concepts in the formulation of the band, the valence bond, the crystal field and the structural ligand theories, the last mentioned being an introduction to the prosthetic group in enzymes, to interpret the mechanism of the electron shift implicit in the chemical reaction. We are indebted to G. M. Schwab for formulating these mechanisms on metals and to Wagner, Garner and F. F. Vol'kenstein for its application to semiconductors. The transition from co-operative to individual catalytic units in a solid catalyst has likewise been proposed as a result of extensive work on semiconducting oxides and sulphides.

In spite of many careful experiments, our knowledge of the heats of adsorption especially for chemisorption leaves many questions unsolved. This is also the case with respect to the problem of surface mobility, an understanding of which is fundamental to any theory of mechanism in surface reactions. Concerning recent discussions on heterogeneous catalysis, there emerge from one's memory some remarks made by Eddington to the effect that progress is not measured by the problems which have been solved but by the questions one can ask. In the field of heterogeneous catalysis many old questions still remain unanswered.

Chapter 2

General Properties of Catalysts

Catalysis is the phenomenon in which one or more substances isothermally (and non-radioactively) augment the rate of a chemical reaction without appearing in the stoichiometric equation of the reaction. In heterogeneous catalysis the catalyst constitutes one or more phases which are distinct from those formed by reactants and products.

In cases where the rate is augmented with increase of concentration of one or more products or of the surface of one or more phases formed by products, the phenomenon is called autocatalysis. This definition of catalysis differs only in preciseness from that of Berzelius, who originated the term.

In dealing with reversible processes, e.g.

$$\mathrm{A+B \rightleftharpoons C+D},$$

the intrusion of a catalyst into the system provides an alternative path, but does not alter the equilibrium. Consequently, both forward and reverse reactions are catalysed to equal extents. It is thus important in investigations on catalytic activity to have prior knowledge of the conditions of equilibrium of the systems under investigation. This is especially the case when we are interested in the problem of yields in the conversion of reactants into products. The small yield of ammonia from nitrogen and hydrogen obtainable under ordinary pressures and high temperatures is a historically important case. In the above reaction with gaseous reactants the free energy relationships can be obtained from thermal data, making use of the Van't Hoff isochore and the Nernst heat theorem

$$\Delta G_0 = -RT\ln K_p = -RT\ln\frac{[\mathrm{A}]\,[\mathrm{B}]}{[\mathrm{C}]\,[\mathrm{D}]}$$

and

$$\ln K_p = -\frac{\Delta H_0}{RT}+\frac{\sum \nu C_p \ln T}{R}+\frac{\sum \nu_\beta}{R}+\sum \nu i$$

necessary specific heats and chemical constants being derived from experimental data.

Although the phenomenon of catalysis can intrude at the liquid–gas or liquid–liquid interface it is with solid surfaces that we are most concerned. Solid surfaces are more complex than liquid surfaces in that they usually present areas or points of non-uniformity whereas the latter are surfaces of equipotential. Many catalytic reactions are brought about by the presence of metals or simple compounds such as oxides or sulphides, and in many cases the admixture of relatively small quantities of another compound can produce a great enhancement of the catalytic activity. The use of such "promoted" catalysts is now general. Since catalytic reactions depend on the available surface, it is common practice to impregnate a carrier or support material with the catalyst and thus produce an extension of the surface. Such carriers may have other important properties: thus they can affect the mechanical strength, impose a restraint on the rate of recrystallization and grain growth or improve the packing characteristics, and, in addition, they may affect the life as well as exerting some promoter effects on the catalyst itself. The promoter effects may be due to epitactic alteration in the atomic spacing of the catalyst or to modification of the "valency" because of incorporation in the crystal field. Since the extension of the catalyst surface is of importance, factors such as porosity, i.e. the size and distribution of the pores and their stability, have to be taken into consideration. It is evident that limitations are to be expected, dependent both on the sizes of the molecules of reactants and products as well as on diffusion rates operative during reaction.

We can thus distinguish between the available surface, the true surface and the apparent surface, i.e. that determined by consideration of the external dimensions of the catalyst alone. The ratio of the available to the apparent surface, sometimes termed the specific surface, consequently depends on the nature of the reactants involved, and the relationship to the true surface ($m^2\ g^{-1}$), in its turn, depends on the method adopted for its measurement.

It is customary to employ the Brunauer-Emmett-Teller (B.E.T.) method for the measurement of the specific surface, using either nitrogen or a rare gas. Although this gives some general concept as to the difference between catalysts prepared in different ways, owing to the complexity of the surface structure and the possibility that catalytic activity is limited to certain regions of the surface, the inferences drawn from B.E.T. data have to be used with caution. This is evident when supported catalysts are under consideration. A gas selectively adsorbed by the catalyst and not by the support should be chosen. For this purpose it is preferable to use either hydrogen or carbon monoxide; the former gas may introduce considerable errors by penetration and solution in the catalyst. Although the chemisorption of carbon monoxide introduces considerable selectivity, there is still some uncertainty, due to the fact that there are two forms of the chemisorbed gas: the linear form and the bidentate form, CO_a— and $CO_a\!\!<$ respectively. There is a high probability that, when a monolayer is complete, only the former (yielding $1{\cdot}2 \times 10^{15}$ molecules cm^{-2}) is present. A strong case can be made for the use of oxygen at 80° K.

The use of radioactive and isotopic methods for the evaluation of the specific surface is becoming more frequent. Careful work on exchange methods of silver and the halogens, where the adsorption takes place by an exchange mechanism, e.g.

$$AgCl + Cl_2^* \rightarrow AgCl_3^* \rightarrow AgCl^* + Cl_2,$$

shows that, although surface coverage is rapid and limited only by diffusion in the gaseous or liquid phase, this is followed by a slower process in which bulk exchange diffusion takes place following the usual Fick law which, for unit surface area, is

$$w = \frac{\partial c}{\partial t} = D\frac{\partial^2 c}{\partial x^2} = 2k\sqrt{\left(\frac{Dt}{\pi}\right)}$$

where D is the diffusion coefficient. Provided that the energy of activation for the diffusion rate is not too low, e.g. $E_D > 1{\cdot}0$ eV, the two processes can be experimentally differentiated from one another.

Another general method for determining the surface area of a catalyst without reference to the specific cross-sectional area of the adsorbate consists in measuring the heat of immersion of the solid in two stages, firstly in the vapour of a liquid up to the saturated vapour pressure and then the additional heat liberated on immersion of the vapour-saturated solid in the liquid. We obtain

$$\Delta H_{sl} = \Delta H_{sv} + \Delta H_{s*l}$$

where ΔH_{sl} and ΔH_{sv} correspond to the heats of immersion in liquid and vapour respectively, and ΔH_{s*l} is the energy change associated with bringing the wet solid and liquid or two liquid surfaces together. This is equal to $2U$ where U is defined by the equation

$$\gamma = U + T\frac{\partial \gamma}{\partial T}$$

γ being the free surface energy of the liquid covering the adsorbent.

Evidently, this method is not suitable for exact work because parts of "accessible" surfaces may fill up with liquid in the capillaries as the vapour pressure increases in accordance with the Kelvin equation and, in addition, even on the plane surfaces, as the vapour pressure p approaches p_0, more than a monolayer of vapour will be condensed, liberating heat.

As might be anticipated, since the surface of the solid catalyst alone is involved in the catalytic process, catalysts are extremely sensitive to the presence of impurities, which can find their way to the surface either from the surrounding gaseous or liquid phase, or by a process of diffusion from the interior of the catalyst or its support. Such impurities can be adsorbed on the surface and thus act as poisons.

In many cases poisons can be removed unchanged by desorption. Other means of removal are as a compound formed with the catalyst by suitable elevation of the temperature and by displacement by another gas or conversion into a more volatile product by chemical reaction, frequently with air or steam. If the quantity of poison in the reactants, or formed as a side reaction during reaction, remains constant, and the poison is a reversible one, the reaction rate may acquire a steady state after a period of induction. The rate under these conditions may bear no relationship to the true catalytic activity.

In experimental work with metallic catalysts, reactions can be studied with the catalyst in the form of a wire that can be electrically heated to any desired temperature. This method, due to Irving Langmuir, has a wide range of applicability. Equally important is that involving the evaporation of the catalyst and condensation as a film or mirror on the interior surface of a cooled container. Evaporation can be effected as originally employed by O. Beeck by means of a hot wire, or by electron bombardment of a small bead of the metal. Coprecipitation of a mixture of soluble salts of metal and carrier permit the preparation of active supported catalysts. Metals can also be deposited on carriers such as magnesia, magnesite or diatomite brick either directly or by ignition of salts deposited by evaporation from a solution; the most suitable salts are those that leave no appreciable residue except the metallic oxide. Oxalates, carbonates, formates or nitrates are customarily employed, the oxide being subsequently reduced by means of hydrogen.

Amongst the methods which have more limited application are the electrolytic deposition of the catalyst on a less active electrode, the deposition by cathodic sputtering or by condensation of the vapour from a vapour gun.

Highly porous metallic catalysts may be formed from alloys of nickel or cobalt with aluminium or silicon, the metallic skeleton derived by leaching out with caustic soda being a particularly robust and effective catalyst for hydrogenation reactions.

Colloidal solutions of catalytic metals are readily prepared by reduction in solution of the appropriate salts. Their high catalytic activity and many similarities to the biocatalysts, the enzymes, have maintained interest in these somewhat unstable and non-robust systems.

Surface Structure

Crystals grow with well-defined facets in which the lattice spacings are characteristic of the planes. If we regard the lattice energy of a crystal as comprising the sum of the interactions between one atom and its nearest neighbour and next nearest neighbours, each interaction of each type being identical, the energy of adsorption of the atom will evidently depend on its position in the crystal and on the crystal structure. If we denote the two interaction energies by u_1 and u_2, respectively, the energies of adsorption on the different facets of a body-centred cubic crystal will be $4u_1+u_2$ on the [100] plane, $2u_1+2u_2$ on the [110] plane and $4u_1+3u_2$ on the [111] plane. The heat of sublimation $(\Delta H_s+(RT/2))$ is $4u_1+3u_2$. Not only will the edges and corners of the crystallites differ in their free energies as well as the various planes but the spiral growth of crystals will, as Frank has shown, produce screw dislocations; consequently, steps and other surface imperfections (lines of dislocation, stacking faults and twins revealed by diffraction patterns and electron microscopic examination) as well as point defects, i.e. atomic vacancies, will be present. In addition, there will be a variable quantity of intercrystalline cement or amorphous metal. The heat of adsorption on a step is some twice and in a kink some three times that on the plane surface. Although it is possible that some reactions, e.g. the H_2–D_2, reaction, proceed on all the exposed surface of the catalyst, the experimental work available at present strongly suggests that, in general, catalytic activity is confined to limited areas of the catalytic surface.

Active Patches

The active patch hypothesis of H. S. Taylor and the necessity for suitable atomic spacing of the substrate metallic atoms insisted upon by Balandin each has its protagonists. In view of the crystalline nature of the metals, the concept of a Gaussian distribution of patches or assemblies of different free energies is unlikely and it is much more likely that a limited number only of such individual areas exists, namely the different crystal planes that may be present. Again, since the atomic spacings on different planes may differ, it is most likely that the catalytic activity may vary from plane to plane or even be confined to one plane. On each of these planes there must exist lattice vacancies and dislocations. The extent of these in a metal can be determined by the rate of release of strain energy, and by changes in density, electric resistivity or hardness. In metals such as copper and nickel, changes in these properties take place within narrow temperature ranges. At lower temperatures (T_r), the atomic movement is sufficient to fill up the vacancies and at a higher temperature (T_d) recrystallization removes or diminishes the lattice defects.

The temperature range over which the lattice defects suffer diminution in number is affected by the number of defects present and the extent of the consequent lowering of the strain energy.

It has been claimed that the decomposition of benzene diazonium chloride, the dehydrogenation of ethyl alcohol on cold worked copper and a variety of catalytic actions on copper suffer relatively great diminutions in velocity when the catalysts are heated over the two ranges T_r and T_d.

In light of the facts that in annealing processes the area of any particular plane may well suffer a drastic reduction and that in the release of strain energy the interatomic distance changes the inference that the active centres for catalytic activity are confined to the lattice defects and dislocations is scarcely justified at present. It seems more reasonable to regard the crystal facet as uniformly active, although stronger adsorption and even some modified chemical activity at the dislocation need not be excluded.

It is worthy of comment that chemical reactivity and phase discontinuities are found to be related in many cases of chemical action. Ostwald and, later, Langmuir pointed out that in cases of dissociation, e.g. of $CaCO_3$, when equilibrium is established the dissociation of the $CaCO_3$ and its reformation from the resulting CaO must be confined to the common interface in order to satisfy the conditions imposed by the phase rule. Faraday was the first to observe the necessity for a discontinuity and its subsequent autocatalytic growth in the dehydration of salts with water of crystallization, and the reduction of various metallic oxides proceeds most rapidly at the metal–metal oxide interface.

Regional catalytic activity has been experimentally observed for the decomposition of ammonia on a monocrystal of copper where the [111] facet was found to be twelve times more active than the [100] plane. A similar selective action has been noted for the decomposition of formaldehyde on the same metal.

The decomposition of carbon monoxide proceeds most rapidly on the [111] planes of both copper and nickel. More evidence of an indirect kind is available to suggest that certain crystal facets only possess either all or a maximum of catalytic activity. With a model surface of uniform lattice the availability of the surface for adsorption evidently depends on the lattice spacing as well as the dimensional relationship to the adsorbate. In the case of hydrogen atoms on a metallic surface the whole of the surface can be covered, but a mercury atom (where Hg—Hg is 3·05 Å), if placed on a lattice point of the catalytic metals where, on the [110] plane, Ni—Ni is 2·48 Å, Pt—Pt is 2·78 Å, Rh—Rh is 2·69 Å, Ru—Ru is 2·65 Å and W—W is 2·74 Å, will overlap three or four of the nearest substrate neighbours, the number depending on the lattice plane considered. These partly shielded atoms may be capable of accommodating hydrogen atoms or some other adsorbate. This factor of surface availability becomes more clearly defined when multipoint contact is considered. The simplest example of this involves two-point attachment of three different types; two of these, in which two atoms of the substrate are involved for each molecule adsorbed, may be exemplified by the adsorption of carbon monoxide on nickel as

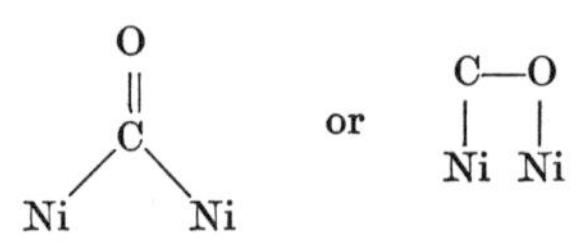

and the other by the adsorption of an olefin on the same metal. An olefin such as ethylene is bonded to the surface either by π- or 2σ-bonding, the two carbon atoms being bonded to two neighbouring nickel atoms. In a face-centred cubic nickel crystal the distance of closest packing is 2·48 Å and the size of the unit cell 3·50 Å. Taking the Ni—C and the C—C distance bond lengths as 1·82 and 1·53 Å respectively, it is not possible to fit ethylene on to the 3·50 Å spaced nickel atoms but only on the [110] 2·48 Å spaced plane. If a tetrahedral angle between carbon and metal atom is assumed, the metal spacing would be some 2·73 Å suggesting that rhenium (2·74 Å) and rhodium (2·69 Å) would be the most active metals; if the lattice expands from palladium, platinum to iron, the activity falls and, if the lattice contracts to nickel and tantalum, the activity is likewise reduced. In accordance with Balandin's views, the hydrogenation of benzene on alloys of iron and nickel takes place only on face-centred cubic structures, where it is possible to get adsorption of the benzene ring as a π-adsorption complex on a hexagonal lattice. Ernest and Skau extended this work to copper, nickel, silver and palladium alloys. Investigating the silver–palladium system, the palladium, being face centred and within the Balandin limits, was active but the silver, which is also face centred but outside the Balandin limits, was inactive. There is a decrease in activity in the alloys as the proportion of silver is increased. This decreased inactivity in hydrogenation may be in part, as we shall have occasion to note, a mere reflection of filling up the d-band vacancies, the catalytic metals themselves possessing d-band vacancies. Superimposed on this is the necessity for the presence of a hexagonal array of catalytic metallic atoms: thus it has been claimed that small quantities of rhenium on a carbon carrier are more effective than larger quantities deposited on alumina, where the cubic crystalline γ-Al_2O_3 is less effective as a support than α-Al_2O_3.

From models it can be shown that ethylene will almost cover the whole surface but substituted ethylenes, increasing from isobutene to trimethylene, cannot be accommodated in close packing and they leave isolated metal atoms accessible to the adsorption of hydrogen, as revealed by the ortho–para reaction.

The third type of reaction involving two-point attachment is to be found in the adsorption of diatomic gases, where dissociation is involved. Since adsorption takes place in pairs when the lattice has been filled there will remain, if the chemisorbed atoms are not mobile, some 8·6% of the surface bare in the form of isolated sites. Above these sites molecules can be held more tenaciously than above the covered surface. We shall note that several varieties of adsorbed species may co-exist even for the elementary gases.

Physical Adsorption

Physical adsorption or van der Waals adsorption, in contrast to chemisorption where an electron switch is involved, is reserved for those cases where the dispersive forces are operative. The dispersive forces are those that exist between all molecules and which are due solely to quantum mechanical resonance of the electronic structure of the molecules involved. Even the noble gas atoms with spherical symmetry over a time average develop non-symmetric configurations due to the electronic zero-point motion. Evidence of this is the fact that the evaporation of a liquid or solid requires the expenditure of energy. This force is responsible for cohesion. Contact between two phases of different cohesive energies results in adhesion and such adhesion must exist between solid–vapour, solid–liquid and liquid–liquid interphases. Information as to the nature and magnitude of these forces can be gained from a study of each of these systems.

The calorimetric heats of physical adsorption of gases and of vapours on both high energy and low energy surfaces are comparable to the latent heats of evaporation, or the heats of adsorption are comparable with the cohesive energies of the adsorbates. The Langmuirian concepts of monolayer adsorption are not applicable here since, under conditions of temperature and pressure when adsorption can take place, condensation may also occur to build up a thick layer. The greater the disparity between the adhesion and cohesion the closer will the adsorption isotherm approximate to the Langmuirian form whilst, conversely, the smaller the difference and the closer the temperature to T_c, the two-dimensional critical temperature, the thicker the film. The presence of capillaries in the substrate results in capillary condensation more readily than on the plane surfaces. As a result, the isotherms for physical adsorption may take on various forms when v–P/P_0 plots are made, v being the quantity of adsorbate adsorbed as a liquid and P_0 being the vapour pressure of the liquid. The B.E.T. equation is customarily employed for determining the specific surface. This equation is based upon the premise that the bulk adsorption can be regarded as a series of superimposed monolayers each conforming to the Langmuirian conditions with suitable values for the two independent constants, α and ν, in the equation (Chapter 4). Clearly the premises of equilibrium between evaporation and condensation for each layer in a multilayer system are not tenable and, consequently, we are faced with two problems. Firstly, how thick must the layer be before the free surface acquires the property of the bulk liquid? Since the adhesive forces are contributing to the cohesion of the adsorbed vapour, they evidently extend beyond one molecular diameter but fall off with distance from the adsorbent. Secondly, we must inquire whether the condensed phase is to be regarded as a normal liquid or whether it possesses some structure.

The Potential Theory

Eucken, Williams and Polanyi were the first to develop a potential theory in which an interaction between the adsorbed molecule and the substrate occurred in an adsorption "volume", the adsorbate being regarded as a liquid. There exists, on this view, at any distance x from the surface, a plane of equipotential of adsorption ϵ_x. This potential is in fact the free energy change ΔG in an isothermal transfer of substance in the liquid state to the space in the adsorption volume, or,

$$\epsilon_x = RT \log \frac{P_0}{P_x}$$

If the law of attraction be written in the form

$$\frac{\partial \epsilon}{\partial x} = -\mu A_s^{1/2} A_v^{1/2} \phi(x)$$

where A_s and A_v are the cohesions of substrate and vapour respectively, that for the vapour being obtained from the van der Waals equation, we obtain

$$\epsilon = \int_{\infty}^{x_0} \mu A_s^{1/2} A_v^{1/2} \phi'(x)\,\mathrm{d}x = \mu A_s^{1/2} A_v^{1/2} \phi(x_0)$$

where x_0 is the depth of the adsorption space. The adsorption potential can be plotted against the weight of gas taken up to give characteristic affinity curves, and these should for any given substrate be dependent only on the square root of the cohesion of the gas. Each isopotential layer can be regarded as having a finite volume $\phi(x)$ and the total volume of the adsorption space is regarded as constant and at saturation to hold the same volume of liquid whatever its nature, i.e. $W_x/\Delta = \text{constant}$ where W_x is the weight of vapour adsorbed and Δ the density of the vapour as liquid at the temperature of measurement. Consequently, if the adsorbed liquid is regarded as incompressible $\phi(x) = W_x/\Delta = W_x\bar{v}$ where $\bar{v}$ is the molecular volume. Further, for any one value of ϵ_x the ratio of the amounts of vapour adsorbed at two different temperatures is equal to the ratio of the densities of the adsorbate at those temperatures. Experimental data on adsorption go far to support these views but, as Dubinin has pointed out, this hypothesis cannot be valid to include both surface and capillary condensation. On Polanyi's view the adsorption potential ϵ or ΔG is assumed to be temperature invariant, i.e.

$$\partial\left(\frac{\Delta G}{\partial T}\right)_{x,\bar{v}} = 0$$

Under these conditions the entropy of adsorption must be negative in sign, but the differential molar entropy of vapour adsorption $\Delta S = \partial(\Delta G/\partial T)_x$ is positive for multilayer adsorption. Taking the Kelvin equation

$$\Delta G = RT \log \frac{P_0}{P} = \frac{2\gamma\bar{v}}{r}$$

where γ is the surface tension of the liquid which wets the capillary and $\bar{v}$ the molecular volume at a constant degree of filling, we obtain

$$\partial\left(\frac{\Delta G}{\partial T}\right)_{x,\bar{v}} = \frac{2}{r}\left(\frac{\partial \gamma\bar{v}}{\partial T}\right)_{x,\bar{v}}$$

which must be negative in sign and in consequence ΔS is negative. We may conclude that Kelvin capillary condensation is incompatible with the general potential concept which only sets in over the region where the former relationship breaks down, i.e. for small values of r or when the entropy of adsorption changes from a negative to a positive value. The limiting size of a Kelvin capillary is thus ~ 14 Å. Dubinin has developed a general expression for vapour adsorption on porous solids in the form

$$W_{x\bar{v}} = V_0 \exp - \frac{BT^2}{A_v^2} \log \frac{2P_0}{P}$$

where V_0 is the micropore volume, B an intrinsic property of the porous substance and, A_v the affinity of the vapour. Taking some standard substance as $A_v = 1$ since $\Delta G \propto A_v x\bar{v}$, A_v for any substance can be determined from the ratio $A_v = \Delta G_1/\Delta G_2$.

That the substrate may be porous in character draws our attention to the fact that, on exposure of a solid to an adsorbable vapour, the linear dimensions of the solid undergo an extension. The surface tension of a solid is not identical with its surface free energy. On extension of a liquid surface, fresh molecules from the interior bulk liquid enter the surface phase to fill up the voids caused by the extension. This, however, is not possible in the case of solids and thus a separate stress is created, and the change in ΔG (the surface free energy) that takes place when a solid swells on adsorption of a vapour is thus not $\int d\gamma$ but rather $\int d\gamma -$ (work of expansion) which latter may be of the same magnitude as $\int d\gamma$.

The decrease in free energy and the expansion of a solid when vapour is adsorbed were attributed by Bangham and his co-workers to a surface pressure acting against the contractile or cohesive surface forces.

The adsorbed vapours were regarded as a two-dimensional gas obeying the van der Waals equation of state

$$\left(\pi + \frac{\alpha}{A^2}\right)(A - A_0) = kT$$

where π is the surface or spreading pressure which in turn may be derived from the vapour pressure with the aid of the Gibbs adsorption isotherm

$$d\pi = -d\gamma = \Gamma \, d\mu = \Gamma RT \, d \log P$$

We shall have occasion to observe that surface mobility in physically adsorbed gas is by no means a general phenomenon.

If the expansion of the solid is proportional to the decrease in the free energy

$$\delta x = -\bar{\lambda}\,\mathrm{d}\gamma$$

the proportionality factor $\bar{\lambda}$ must be related to the elastic constants of the material. Bangham wrote

$$\bar{\lambda} = \frac{100A\rho}{E}$$

where E is Young's modulus and A and ρ the specific surface and density respectively of the adsorbent.

Since the expansion in most cases is isotropic it is preferable to relate the expansion constant $\bar{\lambda}$ to the bulk modulus K rather than to Young's modulus, the expression then becomes

$$\bar{\lambda} = \frac{200A\rho}{9K}$$

Some of the results obtained are given in Table I. The concept that the solid surface is uniformly stressed cannot be applicable except in special

TABLE I

Substrate	Vapour	$\bar{\lambda} \times 10^3$ (cm dyne^{-1})	Sp. surface (cm^2 g^{-1} 10^6)	Young's modulus
Charcoal				
at 65°	$(CH_3)_2O$	3·9	9·74	5·5
(1) at 10°	EtCl	3·9	9·41	5·3
(2) at 10°	EtCl	4·0	9·45	5·2
at 20°	Butane	3·8	9·07	5·3
Coal weakly coking	n-Hexane	29	1·68	0·0085
Coal extracted with methanol	CH_3OH	26	1·86	0·1

cases, since surface cracks are always present and the fracture stresses of the material must be lowered in accordance with the Griffith equation

$$\sigma_f = \sqrt{\left(\frac{2E\gamma}{\pi l}\right)}$$

where l is the length of the surface Griffith crack.

The general evaluation of the potential energy of interaction is frequently accomplished by means of the empirical expression due to Lennard Jones

$$\Delta E = \frac{a}{r^9} - \frac{b}{r^6}$$

where a and b are constants referring to the properties of the adsorbent and adsorbate.

Calculations have been made of the magn˙tude of the dispersive forces acting between molecules, based upon their respective polarizabilities or magnetic susceptibilities. The first of these, due to London, takes the form

$$\Delta E_d = -\frac{3}{2}\frac{\alpha_1 \alpha_2}{(1/h\nu_1)+(1/h\nu_2)}\frac{1}{r^6}$$

where α_1 and α_2 are the polarizabilities and ν_1 and ν_2 are the principal frequencies in the dispersion of the atoms or molecules. A more useful form was developed by Kirkwood and may be written

$$\Delta E_d = -2\left\{\frac{he}{64\sqrt{m}}\right\}\left(\frac{\alpha_1^{3/2}\alpha_2^{3/2}}{(S_1 S_2 Z_1 Z_2)^{1/2}}\right)\frac{N_1 N_2}{(r_1+r_2/2)^2}$$

where h is Planck's constant, e and m the electronic charge and mass and S and Z the sum of the valency and outer electrons respectively in the atom (1) or molecule (2). N_1, N_2 are the number of molecules per cm^3 in each phase and r_1, r_2 their respective radii.

Lastly we have the Kirkwood-Müller relationship

$$\Delta E_d = -6mc^2\frac{\alpha_1 \alpha_2}{(\alpha_1/\chi_1)+(\alpha_2/\chi_2)}\frac{1}{r^6}$$

where c is the velocity of light and χ_1 and χ_2 the magnetic susceptibilities. If in addition the substrate is an ionic crystal, there will be the additional electrostatic energy in any given portion of a polar molecule adsorbed by the surface. This is given by

$$\Delta E_e = -\tfrac{1}{2}\alpha F^2$$

where F is the field strength which for various body-centred ionic crystals was first calculated by Lennard Jones and Dent.

Finally, there is the repulsive potential that maintains the adsorbed molecule in its potential trough at the surface. From quantum mechanical considerations this takes the form

$$\Delta E_r = K\,\mathrm{e}^{-ar}$$

Born and Meyer showed that $a = 1/0{\cdot}35$ and $K = b^+b^-$ where b^+ and b^- were characteristics of the ions involved. Some values of K are: Li^+, 3·96; Na^+, 12·63; K^+, 31·02; Rb^+, 45·87; Cs^+, 67·85; F^-, 24·96; Cl^-,

71·90; Br^-, 103·3; I^-, 176·1; Ar, 36·6. The total potential energy at the surface of an ionic crystal is thus

$$\Delta E = \Delta E_d + \Delta E_e - \Delta E_r$$

which has to be evaluated over a suitable number of neighbours of the adsorbing ion. Calculations for the case of argon on caesium chloride carried out by Orr and later by Young were in good agreement with the experimental values. Kiselev has compared the calorimetric heats of adsorption of hydrocarbons on graphite with those derived by means of the Lennard Jones method of treatment taking various values of r_{eq} and summing over different numbers of substrate atoms. Some of his data are given in Table II.

TABLE II

Hydrocarbon	ΔH_{obs} (kcal mole^{-1})	ΔH_{calc} (kcal mole^{-1})
n-Butane	8·8	8·3
n-Pentane	10·0	10·1
n-Hexane	12·1	12·0
n-Heptane	14·0	13·9
n-Octane	16·0	15·7
Neohexane	10·0	9·0

Padday has recently made use of the Kirkwood expression to calculate the surface tensions (Table III) of some hydrocarbons and of water utilizing the following values for S, Z and α for the CH_3 and CH_2 groups in the hydrocarbons, the surface tensions being half the cohesional values derived from the adhesion between like molecular species.

TABLE III

Substance	γ_{calc} (erg cm^{-2})	γ_{obs} (erg cm^{-2})
Hexane	19·0	19·2
Dodecane	26·2	25·4
Octadecane	28·6	28
Eicosane	28·7	29
Water	19·2	72

S and Z are taken as 6 and 7 for CH_2 and CH_3 groups respectively, $\alpha(CH_2)$ is $1·69 \times 10^{-24}$ and $\alpha(CH_3)$ is $2·05 \times 10^{-24}$. He found that the work of adhesion between water and wax as calculated was 53·6 erg cm^{-2} compared with measured values lying between 52·2 and 56 erg cm^{-2}. It is interesting to observe that although the dispersive forces alone can account for the free surface energies of the hydrocarbons, they contribute only a part of the free surface energy of water but are the principal ones involved in adhesion. This is especially important in seeking an interpretation of the mechanism of wetting. The decrease in free energy of a solid on immersion in a liquid, i.e. the adhesion to and wetting of the solid is given by the equation of Dupré

$$\Delta G_{sl} = \gamma_s + \gamma_l - \gamma_{sl}$$

where the γ's are the respective surface energies. If a drop of liquid be placed on the surface of the solid to make a finite angle of contact, θ, with it, it can be shown by the principle of virtual work that

$$\Delta G_{sl} = \gamma_l(1 + \cos\theta)$$

an equation due to Young. In this form the difficulties associated with the determination of γ_s to which we have referred are disposed of. If θ becomes zero the liquid spreads over the solid. If there were no adhesion of the liquid to the solid θ would be 180° and on increasing the adhesion there is a decrease in the wetting angle which becomes zero when $\Delta G_{sl} = 2\gamma_l$, i.e. the adhesive forces become equal to the cohesive forces of the liquid; but we have seen that it is the cohesive force contribution to the free surface energy of water which alone is involved. If the adsorbing surface is a metal, calculation of the heat of adsorption becomes more difficult since it depends on the assumptions made as to the electronic state of the metal. If it is treated as a covalently linked structure, Kirkwood's method of treatment is applicable. If, on the other hand, it be considered as a uniform sea of electrons, a number of different expressions have been derived of which that due to Lennard Jones is the simplest, viz.

$$\Delta E = -\frac{mc^2}{2Nr^3}\chi$$

Bardeen, treating the metal as well as the adsorbed molecules quantum mechanically, derived

$$\Delta E = -\frac{mc^2}{2Nr^3}\chi\frac{2{\cdot}6e^2/2r_s h\nu}{1+(2{\cdot}6e^2/2r_s h\nu)}$$

where r_s is the radius of the sphere that contains one metal electron and ν is the characteristic frequency of the adsorbed molecule.

Margenau and Pollard considered the metal surface as a number of small attractive elements obtained by evaluating the interaction between the molecule and the dipole field of each element

$$\Delta E_a = -\left(\frac{e^2 h^2}{m}\bar{\alpha}\frac{f}{\nu} + \frac{2\cdot 6 e^2 a}{2r_s}\right)\frac{1}{8r^3}$$

where

$$f = \frac{m\alpha}{e^2}\nu^2 \quad \text{and} \quad \bar{\alpha} = -\frac{\rho e^2}{4\pi^2 m\nu^2}$$

For the repulsive potential of the metal surface we can either assume, with Lennard Jones, that it is a definite fraction of the attractive potential at the equilibrium distance or assume the equilibrium distance and attempt to calculate the repulsive potential developed by a molecule containing electrons. Pollard has attempted this method of attack and has developed a series of expressions from which ΔE_r may be calculated for various distances of separation. Kemball has calculated in this way the repulsive potential of benzene at a mercury surface (Table IV).

TABLE IV

Distance of separation (Å)	12 C—H electrons (kcal mole^{-1})	12 C—C electrons (kcal mole^{-1})	4 π-electrons (kcal mole^{-1})	$\Sigma\, \Delta E_r$ (kcal mole^{-1})
2·0	48·96	22·32	71·6	142·9
2·5	6·685	0·516	10·14	20·34
3·0	1·123	0·011	2·029	3·526
3·5	0·166	—	0·427	0·593
4·0	0·024	—	0·060	0·084
4·5	0·003	—	0·013	0·016

Using a Lennard Jones equation for the attractive forces, he obtained 18·4 kcal mole^{-1} at 2·7 Å as the maximum limit to the energy of adsorption. With the Margenau and Pollard method of approach, he obtained a lower limit of 10·2 kcal mole^{-1} at 2·95 Å, whilst the value derived from surface tension measurement is 15·4 kcal mole^{-1}.

At interfaces there are two other electrostatic forces that may play a part. Thus, if the system consists of two plates immersed in an electrolyte, with electrical double layers on their surfaces, a pressure will be exerted between the plates due to the mutual repulsion as the double layers commence to approach and penetrate one another. Langmuir calculated this pressure and found it in its approximate form to be

$$\pi = 64nkT\beta^2 \mathrm{e}^{-\kappa x}$$

where n is the bulk ionic strength, $1/\kappa$ is the Debye distance, x is the separation of the plates and

$$\beta = \frac{\exp[(ze\psi)/(kT)]}{2} - \frac{1}{\dfrac{\exp[(ze\psi)/(kT)]}{2}+1}$$

z being the ionic valency and ψ the potential at the interface which may be measured as a ξ potential.

Finally, we may note that where one phase consists of a semiconductor and the other either an insulator or semiconductor there will be an electrostatic attractive force between the two phases which may contribute a sensible amount to the adhesive forces.

We have seen that the important forces operative at an interface are the dispersive and those of thermal agitation. In certain cases electrical forces may likewise contribute. Since bulk adsorption of vapours is readily obtained, it is important to consider in more detail the concept of an adsorptive volume, for abrupt transition between "liquid" and "vapour" seems unlikely. Information on this point can be obtained by consideration of the spreading of liquids both on liquid and on solid surfaces as well as by experiments on the adhesion of plates separated by a small distance. The first accurate experiments on this point were carried out by Sir William Hardy. Hardy noted that there are two methods by which a liquid could spread on another liquid. The drop may flatten to form a thick film or may spread as a monolayer over the surface. On a solid a liquid may spread as such or the vapour of the drop may condense on the surface of the surrounding solid. Hardy found, on examining lenses of benzene on water, that the lens must acquire a critical thickness before its surface exhibits the properties of bulk benzene. If attempts were made to reduce its thickness it became unstable and broke up into lenses of the critical thickness in equilibrium with a monolayer.

Even more convincing were his later experiments on adhesion and boundary lubrication, where he showed that active lubricants facilitate slipping even at 1 μm separation. Readily polarizable substances such as acids, bases and esters were found to be the most effective and he concluded that the frictional efficiency increases with the load and that the vertical orientation of the oil increases as we approach the surface, the capacity of the layer for bearing tangential stress rising as the layer thins. Equally significant was the observation that there is a latent period before the friction falls to its lowest value. During this period the process of molecular orientation is being completed. The influence of the substrate was found to affect the boundary phase; thus Hardy found that the tensile strength of a bar of palmitic acid was $c = 449$ g cm^{-2} but, when enclosed in steel at a separation of $x = 0{\cdot}007$ mm, $c = 15{,}803$ g cm^{-2} and at $x = 0{\cdot}0027$ mm $c = 31{,}000$ g cm^{-2}. A plate falling in such liquids is retarded as it approaches a flooded surface and indeed will eventually float in close proximity to the surface, the so-called Leslie phenomenon. This has been re-examined by G. I. Fuchs, who has made a detailed study of the distance of separation as a function of load and composition of the liquid. Hardy termed this orientation in depth "diachysis," the orientated layer being thicker for highly polar oils at low temperatures. Early attempts to confirm this work were not successful, the intrusion of particulate matter being difficult to eliminate. Amongst the more successful early experiments may be mentioned that of Bangham, who showed that water vapour might be deposited on freshly cleaned mica in layers thick enough to show interference colours before the vapour pressure attained the limiting value of P_{sat}. Quantitative measurements and calculations on this process of multimolecular condensation have been made by Metsik. We can also cite the experiments by Lehner, who found that on compression of water vapour in glass over mercury the PV–P curve showed a pronounced curvature before it became linear indicating a removal of water, computed to be some 5 monolayers in thickness, from the gas phase before P_{sat} was attained.

More recently, the careful experiments of Deryagin and his colleagues have shown that the interaction of the dispersive and electrical forces gives rise to a resulting force, Hardy's diachytic force, which he terms a "disjoining" pressure; evidently this may be either positive or negative depending on whether the forces are repulsive or attractive. On thinning of a layer of liquid on a strongly hydrophilic surface, i.e. where $A_{sl} > A_l$, there is a positive disjoining pressure which opposes further thinning and results in a diachytic structure in the film. On hydrophobic surfaces, on the other hand, the disjoining pressure is negative, so when a critical film thickness is reached the liquid recedes from the surface to give discrete drops and to leave a residual monolayer.

Casimir and Hamaker calculated the dispersive force contribution to the attraction between two metal plates separated by a distance x and obtained

$$F = \frac{A}{6\pi x^3}$$

the contribution to the disjoining pressure in a liquid on a solid surface is accordingly

$$P_{\text{disjoining}} = A_{sl} - \frac{A_l}{6\pi x^3}$$

Since A_{sl} is generally assumed to be of the order of $\sqrt{(A_s A_l)}$ evidently $P_{\text{disjoining}}$ is positive if $A_s > A_l$ and negative if $A_l > A_s$.

The Hamaker constant is $\sim 5 \times 10^{-14}$ erg where x is measured in microns, but more recent analysis indicates that retardative effects reduce the effective values of F on increase of the value of x.

We may conclude that in physical adsorption at low temperatures and pressures the adsorption is Langmuirian, i.e. on fixed sites; as the temperature rises, the adsorbed gas becomes two-dimensionally mobile. On increase of pressure beyond the monolayer range the adsorption proceeds to form a thick film which, for polar systems, possesses vertical molecular orientation attaining the density of a liquid, the adhesional forces being anisotropic.

Activated Adsorption

Like the majority of chemical reactions, the formation of a chemisorption complex and an electronic re-arrangement may involve an energy of activation. This would necessitate a temperature coefficient for the rate of formation of the surface phase whether it be formed by direct impact from the gas phase or from a reactant in a physically adsorbed state. Langmuir pointed out that a clean metal surface could be regarded as possessing free valencies, and that reactions with the surface are comparable to reactions with free radicals which involve no or but little energy of activation. Impurities on the surface which can be displaced by the reactant may induce a slower process, and again an apparent energy of activation may be involved. In contrast to metals, the surfaces of oxides and sulphides on reaction with hydrogen to form chemisorbed hydroxide or sulphydryl layers may be regarded as having energies of activation, and the reactions of gases with chemisorbed monolayers may also show a temperature dependence which may be ascribed to an energy of activation.

Conditions for Conversion of the Physically Adsorbed to the Chemisorbed State

The conditions for condensation and the conversion of physically adsorbed atoms to the chemically bound state can be described most conveniently by means of potential–distance curves (Fig. 1) as indicated first by Polanyi and Lennard Jones.

In the Morse potential curves for hydrogen at a metal surface, r_1 and r_2 are the equilibrium distances of the chemically and physically bound species from the surface through a plane containing the centres of the outermost array of metal atoms and ΔH_c and ΔH_p are the respective heats of adsorption, in this case per mole of hydrogen molecules. The energy of activation E for conversion of the physically adsorbed molecules into chemisorbed atoms is evidently dependent on the magnitudes of ΔH_c and ΔH_p as well as the distances r_1 and r_2. It must be negligible for M—H hydrogen where there is a strong bond as on the transition

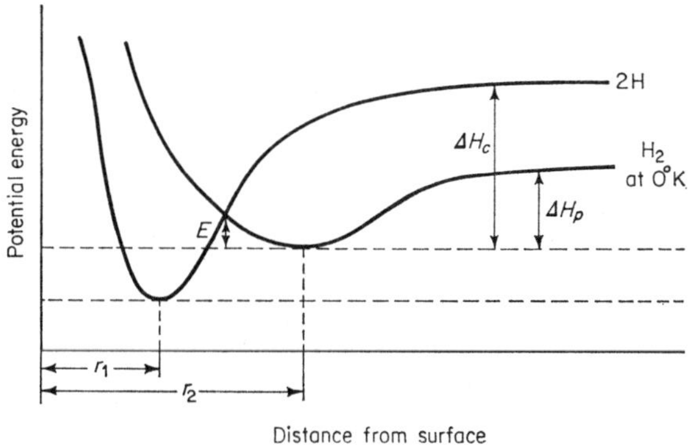

FIG. 1.

metals containing *dsp*-orbitals, but may be present in *sp*-metals such as silver and gold. The probability of transfer of an incident molecule to the bound state is related to the depth of the potential minimum and its own kinetic energy. Only a small quantum of energy sufficient to lower the kinetic energy of the incident molecule and convert it into a highly vibrational state which can exchange with that of the substrate will ensure condensation.

Chemisorption

Chemisorption is the result of a chemical reaction of the adsorbate with the adsorbent and this involves some electronic re-arrangement.

We are indebted to Schwab and especially Dowden for developing this concept in interpreting the action of metallic catalysts, and to Vol'kenstein for its application to semiconductors.

The resulting chemical bond may be ionic or covalent in character and the adsorbed atom or molecule may be regarded as an adsorbed ion, or as an adsorbed dipole in which the orientation may be either positive or negative in direction in relation to the substrate, dependent upon the nature of the chemical reaction. The electrons involved may on one view be those that reside in definite energy levels in the whole metallic crystal, and Dowden has stressed the importance in catalysis of the partly filled *d*-levels in the transition metals and, in some cases, of the overlap of the electrons in the *s*-conductance band with those in the *d*-level. The alternative view, for which there is now increasing support, confines the electronic re-arrangement to those one- or two-electron bonds that hold a specific atom in the crystal. In the catalytic metals the *dsp*-electrons are involved and the overlapping orbitals form the resonating *dsp*-bond. In addition to these bonding orbitals there are, on Pauling's view, atomic non-bonding *d*-orbitals as well as metallic *d*-orbitals present, the former being responsible for any magnetic properties.

In metals the particular surface atom under consideration is fixed in the crystal lattice by ligands to similar atoms, but in catalytic oxides and sulphides the ligands are between the metal ion and unlike atoms; thus the concept of the crystal field as the important factor in catalytic activity is a natural consequence of the valence bond hypothesis for catalytic metals.

If the catalytic activity of a metal or metallic ion is governed by the nature and disposition of the ligands in a crystal, similar dispositions may be found in organometallic molecules, and homogeneous in contrast to heterogeneous catalysts may well result.

The process of chemisorption is most readily exemplified in the case of an alkali metal adsorbed on a metal substrate. We can regard the process as a simple electron transfer from the alkali metal, e.g. potassium to a tungsten substrate, resulting in the formation of adsorbed positive potassium ions that will not only exert an image force of strength $e^2/4r^2$, where r is the radius of the ion or the distance of the ion centre in its potential trough to the surface, but will also build up an electrical double layer, the outer surface being positively charged. The work required to extract a conductance electron from a metal, the electrical work function Φ is made up of two parts, the intrinsic work function ϕ and the Fermi energy E_f of the electrons in the conductance bands (Fig. 2).

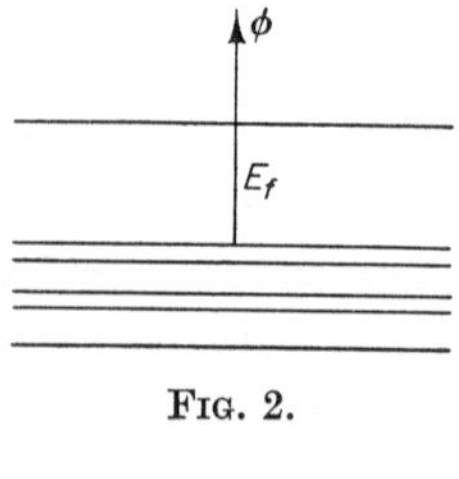

FIG. 2.

Thus

$$\Phi = \phi + E_f$$

Since the height of the Fermi level is not constant but depends upon both the completeness or otherwise of the discrete energy bands in the metal and on the magnitude and size of any double layer formed by adsorption on the surface of the metal, the operative electron work function Φ, i.e. the work required to extract a conduction electron from the metal, is not constant, as the process of adsorption proceeds, although the intrinsic work function ϕ is. The work done in removing an electron from a metal in which the image forces alone are operative is $W = e^2/2x_0$; this is equal to ϕe when ϕ is the work function of the clean metal.

If the shortest distance between adjacent atoms in the surface be x we find

$$x = 1{\cdot}38 \times 10^{-8}\, V^{1/3}$$

where V is the atomic volume and, following Langmuir,

$$\frac{x_0}{x} = 0{\cdot}56 + \frac{0{\cdot}8}{r}\left(1 - \frac{N}{80}\right)^3$$

where r is the maximum valency and N is the atomic number of the substrate. This quantity x_0/x is for most metals $\sim 0{\cdot}60$, rising in the case of the alkalis to $\sim 1{\cdot}0$.

The heat of chemisorption of a potassium atom on tungsten is consequently

$$\Delta H = \Phi - I + \frac{e^2}{4r}$$

where I is ionization potential of the potassium ion, and ΔH must be dependent on the extent of coverage. The change in work function caused by a double layer is given by the equation

$$\Delta\phi = 2\pi n\mu$$

where μ is the value of the dipole moment formed by an adsorbed potassium ion. It may be noted that the Helmholtz relationship $V = 4\pi n\mu$ relates to discrete dipole arrays.

By measuring the surface coverage and the contact potential simultaneously, it was shown by Langmuir for caesium on tungsten and by Bosworth and Rideal for potassium on the same substrate that the value of μ does not remain constant but, as the field changes with the accumulation of ions on the surface as revealed by a lowering of the work function, the dipoles become polarized and, as they are deformable, their apparent dipole moment falls (Table V).

If we regard the surface as containing a mixture of adions and adatoms, the ratio decreasing with increasing concentration, we obtain $r = \sigma_{\text{adatoms}}/\sigma_{\text{adions}}$. Table V exemplifies this. In other words for a close packed monolayer of potassium on tungsten about one atom in five is ionized.

TABLE V

System	Fractional coverage θ	μ (debyes)	$r = \dfrac{\sigma_{\text{adatoms}}}{\sigma_{\text{adions}}}$
Cs on W	0	14·6	
	0·5	6·1	
K on W	0·012	23	0
	0·025	21	0·05
	0·10	13·0	0·6
	0·50	6·0	1·9
	1·0	2·8	4·4

The adsorbed species will evidently evaporate as an atom as long as the energy required to evaporate it as an ion exceeds that for an atom, i.e.

$$\lambda_p + \phi > \lambda_a + I$$

where λ_p and λ_a are the heats of evaporation as positive ion and atom respectively, ϕ is the electron work function of the substrate and I is the ionization potential of the gaseous ion.

For potassium on tungsten we have $\phi = 4{\cdot}6$ eV, $I = 4{\cdot}3$ eV, $\lambda_p(\theta = 0) = 2{\cdot}43$ eV and $\lambda_a(\theta = 0) = 2{\cdot}3$ eV. Thus, for evaporation as positive ions,

$$2{\cdot}43 + 4{\cdot}6 = 7{\cdot}03 > 2{\cdot}3 + 4{\cdot}3 = 6{\cdot}6 \text{ eV}$$

The Boltzmann distribution between ions and atoms is accordingly

$$\frac{\sigma_a}{\sigma_p} = \exp\left(\frac{0{\cdot}43 \times 11{,}600}{T}\right)$$

In consequence, as potassium accumulates on the surface of tungsten the work function falls and, when it falls below $\phi = 4{\cdot}2$ eV, the ratio of atoms to ions rapidly becomes inverted.

That double layers are indeed formed where gases are adsorbed on metals has been established by several different methods: by thermionic measurement of the work function; by photoelectric determination of the work function, with a diode and by measuring the change in capacity of a condenser formed by two plates of the metal held close together when the gas is adsorbed on the plates: methods used with great success by Mignolet and Tompkins. These methods give consistent results in that some gases on adsorption give rise to a progressive increase in the work function and others to a decrease.

As might be anticipated, all the alkali metals form dipoles with an orientation K^+W^-. The same is true, as shown by the extensive investigations of Mignolet, for physically adsorbed gases such as xenon, which has surface potentials ranging from 0·21 V for Xe^+Zn^- to 0·95 V for Xe^+Cr^-. Here we are dealing with induced dipoles in the polarizable noble gases. In view of the strong electron affinity of oxygen ($\sim$6·0 eV) it is not surprising that the chemisorption of this gas results in the formation of a negative dipole M^+O^-. Some values for the surface potentials of adsorbed oxygen on various metals are given in Table VI.

TABLE VI

Metal	W	Ni	Pt	Cu	Ag
Surface potential (V)	−1·90	−1·5	−1·2	−0·68	−0·60

There is a good deal of variation in the published values for the surface potential at $\theta = 1$ for oxygen. This is in some measure due to the nature of the substrate, since different crystal facets have different atomic spacings and work functions. For tungsten the values in Table VII have been recorded, and we might anticipate that oxygen would be preferentially adsorbed on the [111] and [116] planes.

TABLE VII

Facet	111	112	116	001	110	011
ϕ (V)	4·39	4·69	4·39	4·56	4·68	4·53

On metals where low temperature oxidation results in the formation of a chemisorbed monolayer the heat of adsorption falls as $\theta \rightarrow 1$ but if an oxide is formed as a separate immobile phase, e.g. PtO_2, RhO_2, RuO, this phase can increase without any further effect on the levels of the underlying electrons, and the heat of formation is constant.

We shall have occasion to observe that chemisorbed hydrogen can exist in more than one form, a fact reflected in its chemical reactivity. This is also to be noted in the changes in the surface potential of metals when exposed to hydrogen gas. It has been claimed that the surface potentials of copper, silver and gold do not change appreciably on admission of this gas, but atomic hydrogen gives the following values at $\theta = 1$: $Au^{+}H^{-}$, $-0{\cdot}18$ V; $Ag^{+}H^{-}$, $-0{\cdot}40$ V; $Cu^{+}H^{-}$, $-0{\cdot}34$ V; thus confirming the view that dissociative chemisorption of hydrogen does not take place on these metals at ordinary temperatures. Although we would anticipate from the behaviour of hydrogen chemisorbed on oxides that it should be an electron donor, on metals, including the transition metals, the sign of the dipole is, at any rate for small coverages, always $M^{+}H^{-}$. On many metals at high coverages the sign may reverse to give $M^{-}H^{+}$. Some of the negative values recorded are as follows: Pt^{+}–H^{-}, $-0{\cdot}15$; Ni^{+}–H^{-}, $-0{\cdot}35$; Co^{+}–H^{-}, $-0{\cdot}30$; Fe^{+}–H^{-}, $0{\cdot}42$; W^{+}–H^{-}, $-0{\cdot}50$. Positive values for the surface potentials at higher coverages have been recorded for platinum, nickel, tungsten and silver.

To account for the negative dipole of the M—H bond two suggestions have been made: that the covalent bond does represent the formation of hydride hydrogen, or that the hydrogen is in fact an electron donor but that the initial coverage is below and between the outer double layer on the metal; this submerged dipolar array contributes the negative surface potential. When this layer is completed more hydrogen is chemisorbed as $M^{-}H^{+}$ on the conventional surface, and this in consequence reduces the surface potential which may acquire positive values. These two forms of chemisorbed hydrogen can be identified by their different stabilities. Alternatively, electrons may be drawn from the conductance band of the metal to form covalent bonds with the hydrogen; the work function of the metal is consequently increased because of the depression of the Fermi level, the positive dipole at high surface coverages being attributed to H_2^{+}.

Carbon monoxide, another electron donor, does in fact give positive surface potentials on copper ($+0{\cdot}30$ V), silver ($+0{\cdot}31$ V) and gold ($+0{\cdot}92$ V), but negative values on iron ($-1{\cdot}15$ V), cobalt ($-0{\cdot}26$ V) and nickel ($-0{\cdot}46$ V). Nitrogen likewise gives negative values: W^{+}–N^{-}, $-0{\cdot}6$ V; Ta^{+}–N^{-}, $-0{\cdot}38$ V.

The metallic substrate can act as a source or sink for electrons to yield a resulting dipole array electronegative or electropositive in character, the preponderance of either form is dependent on the electron affinity of the adsorbate and on the work function and height of the Fermi level of the metal.

In considering the activation for catalysis of a sequence of gases such as oxygen, carbon monoxide, olefins and hydrogen, we find a transition from strong electron acceptors to electron donors. Evidently all metals with their free electrons should react with strong electron acceptors, e.g. oxygen. Metals with *s*-, *p*-bands should make chemisorbed complexes with carbon monoxide and with olefins, but the electron concentration in the metal must be greatly reduced to render it capable of chemisorbing hydrogen or other electron donors. Such metals have, in general, vacancies in their *d*-bands. Conversely, as the electron concentration or the "valency surplus" of the metal increases, the energy of activation for the electron transfer from adsorbate to adsorbent increases; consequently, good hydrogenating catalysts and those capable of bring about the o–p hydrogen reaction are to be found only amongst the metals containing incomplete *d*-bands, where the energy of activation for transfer of the electron from the gas to the substrate would be small. Table VIII illustrates the behaviour of nickel with a reacting gas.

TABLE VIII

Gas adsorbed	O_2	CO	N_2O	H_2O	H_2
Direction of electron transfer	↑	↑	↑	↓	↓
Change in conductivity	−	−	−	+	+
Change in magnetic property	+	+	+	−	−

The transition metals are characterized by a narrow *d*-band with many levels overlapped by a broad *s*-band, and both these bands are only partially filled with electrons. In the transition metals there are several possible electron configurations, the most suitable wave function is one where a linear combination of the separate states give a minimum free energy. The electron structure of the nickel atom is

$$1s^2 2s^2 2p^6 3s^2 3p^6 3d^8 4s^2$$

In catalysis we are concerned only with the outer bands of the 3*d*- and 4*s*-atomic levels. The Fermi level is part way up the *d*-band, which is full in the case of copper.

The percentage *d*-character of the catalytic metals on Pauling's view is: V, 35; Cr, 39; Mn, 40·1; Fe, 39·7; Co, 39·5; Ni, 40·0; Mo, 43; Ti, 46; Ru, 50; Rh, 50; Pd, 46; Ta, 39; W, 43; Re, 46; Os, 49; Ir, 49; Pt, 44.

For nickel there are two states of valency

	3*d*	4*s*	4*p*
Ni_a↑↓↑↑	..	.	...
Ni_b↑↓↑↓	...	.	..0

where . represents the bonding orbitals and 0 the empty metallic orbitals. Thus Ni_a has two unpaired electrons and d^3sp^3-bonds whilst Ni_b has no unpaired electrons, d^3sp^2-bonds and an unoccupied (metallic) orbital. The saturation magnetic moment of nickel indicates the presence of 0·6 unpaired electrons and the metal is thus 30 % Ni_a and 70 % Ni_b. The percentage *d*-structure of nickel is then 40 %.

Since there are no vacancies in the *d*-band of copper, this metal should only chemisorb hydrogen with difficulty and thus be relatively catalytically inactive for the ortho–para reaction. It can, however, adsorb atomic hydrogen and can at high temperatures also bring about the ortho–para reaction. Whether this is due to impurities acting as electron acceptors or to the thermal promotion of an electron to the *s*-band has not been clearly established. It may be observed that the energy required for such promotion is some 3·0 eV for copper and 4·0 eV for gold.

Very strong support for Dowden's view that unpaired electrons in the *d*-band and *d*-band vacancies play an important part in these chemisorptive processes, resulting in a covalent dipole, can be obtained from a study of alloy systems. The vacancies in the transition metals *d*-bands can be filled by the addition of copper, silver or gold.

Couper and Ely found for the palladium–gold system that from 100 % Pd to 40 % Pd the energy of activation for the ortho–para conversion was only 3·8 kcal $mole^{-1}$. On further increase of the gold content the energy of activation rose rapidly to 8 kcal $mole^{-1}$ and then climbed steadily to 17·5 kcal $mole^{-1}$ for pure gold. Morris and Schwab found that a 1 % dispersion of nickel in copper was active for the hydrogenation of benzene but, on heating, the nickel became molecularly dispersed, with a resultant loss in catalytic activity. The copper–platinum system has been examined by several investigators. Schäfer noted the parallelism between the change in rate of the dehydrogenation of ethane and the change in the accommodation coefficient both associated with change in the energy in the carbon–carbon bending motion of the ethane molecule as the composition of the alloy changed over the regions 0–60 and 60–100 % Pt.

The energy of activation for the ortho–para conversion has also been found to rise rapidly from 84 % Cu (16 % Pt) to 100 % Cu. On the copper–nickel alloys the heats of adsorption of hydrogen and the energies of activation for the ortho–para conversion are said to be uniform within the range 20–90 % Ni, the catalytic activity rising rapidly over the ranges 0–20 % and 90–100 %. Emmett found that the hydrogenation of ethylene proceeded very slowly on copper but rose rapidly on the addition of nickel, up to 20 %, the activation energy falling at the same time. On further addition of Ni the velocity of conversion and activation energy remained constant until 87·4 % Ni was present. On still further addition of Ni there was again a rise in the catalytic activity. Many other examples of the behaviour of such alloys have been furnished by Schwab and his co-workers and by Dowden and Reynolds.

Eley and Knights have attempted to separate the functions of the two electronic factors, namely the effects of *d*-band vacancies and the alteration in the height of the Fermi level by examining the effects of alloying palladium with gold on the decomposition of nitrous oxide. The reaction rate follows the scheme

$$\omega = \frac{P_{N_2O}}{1+\alpha P_{N_2O}+\beta P_{O_2}}$$

and involves basically a transfer of an electron from and to chemisorbed oxygen in steps

$$N_2O(g) \xrightarrow{k_1} N_2(g)+O_a \qquad (a)$$

$$2O_a+2e \underset{k_3}{\overset{k_2}{\rightleftharpoons}} O_2^{2-} \qquad (b)$$

$$O_2^{2-} \underset{k_4}{\rightleftharpoons} O_2^- + \epsilon \qquad (c)$$

$$O_2^- \underset{k_5}{\rightleftharpoons} O_2(g)+\epsilon \qquad (d)$$

In these reactions, (a) and (d) govern the kinetics. On pure palladium, E is ~ 43 kcal mole^{-1}. The reaction is strongly inhibited by oxygen but not by nitrogen, and it is assumed that when a stationary state obtains on the catalyst surface

$$\frac{\partial \theta_{O_2^{2-}}}{\partial t} = \frac{\partial \theta_{O_2^-}}{\partial t} = \frac{\partial \theta_O}{\partial t} = 0$$

where $\theta_{O_2^{2-}}$, $\theta_{O_2^-}$, θ_O are the fractions of the surface covered with the respective reactants.

The rate of decomposition of the nitrous oxide is

$$= KAP_{NO_2}(1-\theta_{O_2^{2-}}-\theta_{O_2^-}-\theta_O)$$

and the rate of liberation of oxygen is given by

$$\tfrac{1}{2}\omega = \{k_5\,\theta'_{O_2^{2-}}-k_1 P_{O_2}(1-\theta_{O_2^{2-}}-\theta_{O_2^-}-\theta_O)\}A$$

On pure palladium the oxygen is strongly adsorbed, with a heat of adsorption of $\Delta H \sim 32$ kcal mole^{-1}. On the addition of gold, the *d*-band vacancies are progressively filled and the heat of adsorption falls to negligible values at $\sim 50\%$ Au. Here the apparent energy of activation for the decomposition of nitrous oxide falls from 43 kcal mole^{-1} to some 12·8 kcal mole^{-1}. On gold-rich alloys (40–0 % Pd) the velocity of decomposition remains constant but the apparent energy of activation continues to fall from a high value, setting in at 40 % Pd where the *d*-bands are filled, and there is no retardation by oxygen. In this region it is assumed that defect activity plays a dominant role in the reaction and that the subsequent fall in the energy of activation can be ascribed to a rise in the Fermi level which becomes significant over this range of composition. It may be observed that palladium saturated with hydrogen should have all its *d*-band vacancies filled and thus be rendered inert or possess a high energy of activation.

The assumption that these alloys are ideal solid solutions over the whole range of composition is not justified unless experimentally verified. Thus, in the case of the nickel–copper system, the heat of mixing to form the alloy is endothermic, $\Delta H \sim 460$ cal (g atom)$^{-1}$. Although at high temperatures the entropy is sufficiently large to ensure uniform composition, at low temperatures we might anticipate separation into a two-phase system. Sachtler has confirmed the presence of such separation at $\sim$200–300°C. The progressive addition of nickel to copper results in a single phase of nickel in copper up to 18% nickel when a complex of composition 18% Ni–82% Cu is formed and further increase of the nickel content results in a two-phase system consisting of nickel and the nickel–copper complex. Sachtler makes a further important observation that, because of the higher diffusivity of copper, the complex will not be distributed uniformly through the bulk nickel but will tend to accumulate at the surface. It is also possible that the free surface energy of the nickel is lowered by the presence of a capillary active complex. Thus a small quantity of complex will be sufficient to coat the nickel and it is only at high nickel contents that the exposed surface will consist of nickel together with diminishing patches of the complex. On this view the nickel–copper alloys should show three catalytic activity–composition ranges; one from 0 to 18% Ni of changing catalytic activity with temperature, a second one of constant activity from 18% Ni to a very high nickel content and a third of changing activity as the relative area of exposed nickel to that of the complex increases. We may also note that in a binary alloy in general both components may have some catalytic activity and, consequently, on progressive change in the environment of one of the components, we would not expect to find definite "break points" in activity but rather changes in rate which become more marked the greater the contrast in the components. Again it may be observed that, although the transition metals have incomplete *d*-bands in the crystal, as atoms they have incomplete *d*-shells and these are not identical. The values for iron are $3d^6 4s^2$ in the atom and $3d^{7\cdot 8}4s^{0\cdot 2}$ in the crystal; the comparable values for nickel are $3d^8 4s^2$ and $3d^{9\cdot 4}4s^{0\cdot 6}$.

The participation of *d*-orbitals in the formation of the surface covalent bond leads to the view that the greater the % *d*-character as defined by

Pauling the less orbitals there will be available for surface binding of substances like hydrogen. Oxygen, which readily accepts electrons from the s- or p-bands, is not dependent on this characteristic of the metal.

We may write the energetics of formation of a covalent band making use of Pauling's electronegativities, e.g. for $M+H_2 \rightarrow 2MH$

$$\Delta H = 2\Delta H(\text{M—H}) - \Delta H(\text{H—H})$$

$$\Delta H(\text{M—H}) = \tfrac{1}{2}\{\Delta H(\text{M—M}) + \Delta H(\text{H—H})\} + 23{\cdot}06(\chi_{\text{M}} - \chi_{\text{H}})^2$$

where χ_{M} and χ_{H} are the electronegativities of the two reactants.

The value of $\Delta H(\text{M—M})$ is obtained from the heat of evaporation λ of the metal. In a face-centred metal such as nickel there are twelve neighbours to each atom and, for one bond between two atoms, we may write

$$\Delta H(\text{M—M}) = \tfrac{2}{12}\lambda$$

This is also approximately true for a body-centred metal such as tungsten. For the chemisorption of hydrogen on tungsten the value of ΔH calculated in this manner is $\sim 50{\cdot}4$ kcal mole^{-1}, comparable to an experimental value of ~ 45 kcal mole^{-1} at low values of θ.

The electronegativities of the various adsorbates, e.g. χ_{H}, are obtained from Pauling, whilst those of the substrate from their work functions

$$\chi_{\text{M}} = 0{\cdot}355\phi$$

This latter we have seen varies with the extent of coverage and can be corrected for in part by taking into consideration the electrical double layer moment μ.

Since the chemical "bond" is a result of a mutual interaction, the chemical reactivity of the adsorbed species undergoes a change dependent on its polarizability; thus, a cation adsorbed on to a metal surface, an electron sink, results in a species that behaves as if it were in a higher valence state, e.g. K on W behaving as K^+. If adsorbed on an oxide such as Al_2O_3 it behaves as if it were in a state of lower valency.

We note that silver ions adsorbed on alumina acquire the catalytic properties of bulk silver in say the decomposition of hydrogen peroxide, where an electron donor is required. This is found to be the case also with copper, where cupric ions adsorbed on to metallic copper are polarized to an increasingly higher "valence". It is for this reason that the reduction of copper oxide by hydrogen proceeds most rapidly in those regions where electron abstraction from the hydrogen to form protons is easiest, i.e. the cupric ions at the copper oxide–copper interface.

Catalytic Oxides and Sulphides

In the catalytic oxides and sulphides defects may be present. In an ordered system such as

```
A B A B A B
B A B A B A
A B A B A B
```

there is the possibility that regions of disorder may exist such as

```
A B A B A B
B A B B A A
A B A A B B
```

Any increase in entropy favours the disordered state, which must be counterbalanced by the interatomic energies such that AB $\gg$ AA or BB. Apart from these regions of disorder, which in heteropolar ionic systems are still electrically neutral, lattice defects may be present. Defects may be of two types, the so-called Schottky lattice and Frenkel interstitial defects. These may be represented graphically as follows:

```
A B A B A B      A B A B A B
B A – A B A           B
A B A B A B      B A B A B A
B A B A B A      A B A B A B
Schottky defect  Frenkel defect
```

Evidently these are not electrically neutral but must acquire excess positive or negative charges. It is convenient to classify them as electron donating or n-type and electron accepting or p-type semiconductors. Zinc oxide is normally found to acquire n-type characteristics whilst nickel monoxide is usually an electron accepting or p-type semiconductor. Sometimes the concept of positive or negative holes capable of being filled to neutrality is made use of, but electron availability or the height of the Fermi level is frequently found to be a more convenient method of treating such systems.

These two types of semiconductor can be depicted as

$$\frac{e\uparrow}{\text{ZnO}} \qquad \frac{e\downarrow}{\text{NiO}}$$

n-type p-type

We can impart semiconducting properties to neutral zinc oxide, convert it into an n-type semiconductor and raise the Fermi level by forming lattice defects. This can be accomplished either by the addition of excess zinc or, what is equivalent, by the loss of molecular oxygen. The excess electrons may be generated by either reaction (a) or (b).

$$Zn \rightleftharpoons Zn^{+} + e \qquad (a)$$

$$2ZnO \rightleftharpoons 2Zn^{2+} + O_2 + 4e \qquad (b)$$

The first method of representation is more convenient for if we replace a Zn^{2+} ion by doping with a trivalent positive ion such as Ga^{3+} we can regard the effect as equivalent to bringing about the reaction

$$Ga^{3+} + Zn^{+} \rightarrow Ga^{2+} + Zn^{2+}$$

thus causing a diminution in concentration of Zn^{+} in reaction (a), resulting in an increase in the electron concentration in the conduction band, a raising of the Fermi level and strengthening the n-type characteristics or electron donating powers of the semiconductor. Conversely, if we replace a Zn^{2+} ion by a monovalent ion such as Li^{+} we may imagine that by the reaction

$$Li^{+} + Zn^{2+} \rightarrow Li^{2+} + Zn^{+}$$

the Zn^{+} concentration is increased, resulting in a diminution of the electron concentration according to equation (a). This results in a decrease in the electron donating power (or a relative increase in the electron accepting power) or p-type characteristics, and a fall in the Fermi level. Change in temperature will cause a change in the position of equilibrium of these charge generating reactions and thus alter the p- or n-type characteristics. Thus, in Cr_2O_3 the p-type characteristics in an oxidizing atmosphere are converted on elevation of the temperature to 400°C into those of an n-type oxide found in reducing atmospheres. We may note that the effects of doping with traces of ionic cations of different valency may be regarded as equivalent to attributing to the oxide or sulphide the power of inducing a valency change in the foreign or added ion, i.e.

$$Ga^{3+} \rightarrow Ga^{2+}$$

$$Li^{+} \rightarrow Li^{2+}$$

In crystalline materials the distribution of defects will, above the Tamman temperature, be uniform throughout the crystal and at equilibrium be dependent on the oxygen pressure. Below this temperature there may be a non-uniform distribution. In addition if the free surface energy of the crystal surface is lowered by a defect then at equilibrium there will be a Gibbsian surface excess of defects.

The electrons or positive holes act as conducting systems and the surface conductivity may differ from the bulk conductivity due to the presence of n charge carriers, giving a conductivity $\sigma = nev$ where v is the electronic mobility and e the electronic charge.

The energy of activation E for the increase in conductivity with temperature enables us to determine the number of positive holes in the valence band for p-type semiconductors. This is given by the expression

$$n = 5 \times 10^{15} T^{3/2} e^{-E/2kT}$$

The sign of the carrier can be determined from observation of the direction of the Hall effect or from the thermoelectric potential, and the change in valency which we have associated with the semiconducting properties, e.g. zinc oxide reinforced with Ga^{3+} or Li^{+}, has been investigated in detail by Selwood and by Verwey, chiefly by magnetic methods.

The concentration of charge carriers determined from the Hall coefficient is given by the expression

$$n = 0{\cdot}59 i\phi\delta / eV_H r$$

where i is the current, ϕ the magnetic field strength, δ and r the thickness and width of the sample, e the charge on the carrier and V_H the Hall voltage. The number 0·59 is derived from the lattice scattering factor $3\pi/8$. Chon and Prater used this method for determining the changes in carrier concentration on doping zinc oxide. They obtained the values in Table IX. It is clear that, as oxygen is progressively adsorbed on to an n-type oxide such as zinc oxide, the conductance band loses electrons,

TABLE IX

Dope	Dope (ppm)	n (cm^{-3})
None	None	3×10^{16}
In	45	$7{\cdot}2 \times 10^{17}$
Ga	55	$1{\cdot}2 \times 10^{18}$
Li	500	$2{\cdot}5 \times 10^{14}$

with the result that a space charge is built up. The same is true if the acceptor centres are filled by an electron-donating adsorbate on a p-type semiconductor. Consequently, there is a limit set to the adsorption by the action of the space charge. This equilibrium is established when the potential energy of the electrons in the Fermi level of the semiconductor is equal to the potential energy of the electrons in the adsorbate.

We can depict the electron transfer mechanisms and the building of the surface double layer by the potential distribution shown in Fig. 3.

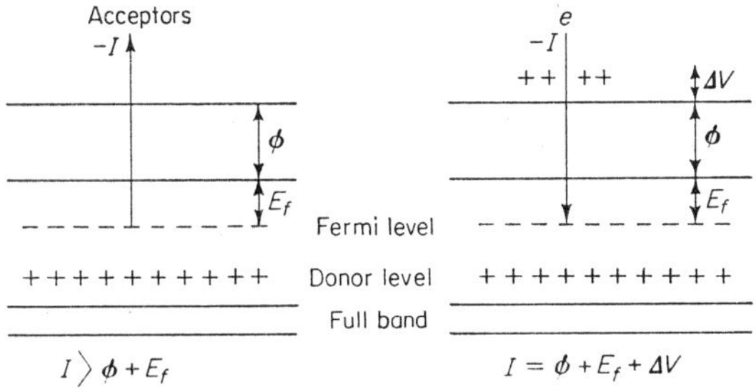

FIG. 3

It is important to pay attention to the state of the catalyst when under actual working conditions, since all these oxide and sulphide catalysts may be regarded as capable of electron donation and electron acceptance in a manner similar to the oxy-redox systems, e.g.

$$Cu^0|Cu^I,\ Cu^I|Cu^{II},\ Cr^{IV}|Cr^V$$

Thus $ZnCr_2O_4$ is normally a p-type semiconductor and when fully oxidized is black; on reduction it is converted into an n-type catalyst and is greenish blue in colour. Even in 90 % CO and 10 % O_2 its electron-accepting powers exceed its n-type characteristics so that it remains fully oxidized.

p-Type catalysts are better than n-type catalysts in an oxidizing atmosphere and n-type catalysts rather than p-type ones should be used in a reducing atmosphere. It is for reasons of this kind that methanol synthesis, performed in a reducing environment, is best effected by n-type catalysts which cut down the chemisorption of carbon monoxide (e.g. ZnO, CuO, Fe_2O_3, V_2O_5), whereas the catalytic decomposition of nitrous oxide, which takes place in a highly oxidizing atmosphere is best carried out on p-type semiconductors, and consequently we find the order of effectiveness of oxides for this reaction in the following order: CoO > Cu_2O > NiO > MgO > CeO_2 > CaO > ZnO > TiO_2 > Fe_2O_3 and p-type oxides > insulators > n-type oxides. That nitrous oxide is an electron donor on adsorption on the cations of the transition metals is confirmed by infrared analysis, where ionic bonding is assigned to $\lambda > 1400$ cm^{-1} and covalent bonding to $\lambda = 1870$–1700 cm^{-1}. On iron the bonding is mainly ionic; on nickel, on the other hand, it is covalent and on chromium no less than five types of bonding, ionic and covalent, have been identified. On the non-transition metallic oxides the gas is adsorbed on the oxygen and not on the metallic site.

We have noted that the catalytic oxides possess p-type or n-type characteristics, but in addition the oxides of the transition metals must often have unpaired *d*-electrons and can, as Dowden has pointed out, form covalent linkages with suitable adsorbates. We may conclude that chemisorption is affected by the rise or fall of the Fermi level, as the *d*-band fills or empties, by the p-type or n-type characteristics of the substrate and electron affinity of the adsorbate and the mutual polarization of the adsorbed dipoles. In semiconductors the p,n-type characteristics are probably more important than the height of the Fermi level since, for example, in silicon the extreme of the former is $\sim 1{\cdot}2$ V but the change in the height of the Fermi level corresponds to only some $0{\cdot}2$ V.

Catalytic metals as well as the oxides and sulphides which we have discussed can be regarded as co-operative catalysts since the electrons required for the chemical change can come from and return to a continuous band. These systems are to be distinguished from those catalysts that rely on a crystal field and those in which one single metallic ion or a pair, by virtue of the relative strengths of their ligand bonding in an organometallic complex, are catalytically active. Recent work suggests that the mechanism of operation of a catalytic ion in a crystalline system is similar to that of the organometallic complexes in that it is the ion and its ligand disposition and strength that comprise the catalytic unit. We shall have occasion to refer to some examples of these catalysts, which may be regarded as transitional to the homogeneous types.

On the Poisoning of Catalysts

The phenomenon of poisoning of metallic catalysts by the adsorption of strongly held extraneous material has long been recognized. Effective poisons for metal catalysts are found in compounds of the elements in Groups V and VI possessing lone electron pairs, the co-ordinate link being established by electron donation to the *d*-band. We can compare the relatively toxic substances SH_2, PH_3, SO_3^{2-}, NH_3, pyridine with the non-toxic PO_4^{3-}, SO_4^{2-}, NH_4^+ and pyridinium$^+$. In examining the poisoning activity of metallic ions on catalytic metals it appears that a minimum of five electrons is required. It has been suggested that the metal ion is tetrahedrally co-ordinated to the metal, inducing a small splitting of the d_γ- and d_ϵ-levels because of the ligand field. Four electrons are found in d_γ-orbitals and the fifth in a d_ϵ-orbital which can take part in a d_π-bonding with the surface atoms.

Several important observations have been made when the quantitative aspects of the effects of the progressive addition of a poison to a reacting system are considered. Firstly, the progressive loss in catalytic activity is not always a linear function of the amount of poison retained by the catalyst. Even a small adsorption relative to the B.E.T. area may cause a drastic reduction in catalytic activity. Again, there are now many well-established cases where selectivity of reactions are to be noted on the progressive addition of a poison. Thus successive additions of CS_2 to a hydrogenating catalyst renders it inactive for the hydrogenation of acetophenone, then for cinnamic acid and nitrobenzene and finally for benzene. Nickel can be poisoned by thiophenone for the hydrogenation of benzene, the catalyst still being active for hydrogenation of phorone. Three sites have been postulated on nickel specific for the hydrogenation of ethylene, benzene and nitrobenzene and respectively poisoned by ethyl sulphide, thiophene and hydrogen sulphide. Thiophene, certain thiocycloalkanes and alkyl sulphides in an atmosphere of hydrogen partially sulphide nickel catalysts at temperatures below 200°C to yield catalysts which hydrogenate diolefins selectively to mono-olefins, whereas mercaptans and disulphides progressively sulphide the nickel and loss of activity occurs, no selectivity being noted.

Poisons may be either reversible or irreversible in character in the sense that reversible poisons include those that allow recovery of catalytic activity on cessation of the admission of the poison. The slow removal or consumption of the poison is revealed as an apparent period of induction. Thus, small quantities of oxygen in an ethylene–hydrogen mixture create with a nickel catalyst an induction period during which the preferentially adsorbed oxygen is catalytically converted to water before the hydrogenation of the ethylene can proceed at an appreciable speed.

The interpretation, long held in the past, that only limited areas of the catalyst are available for catalytic activity and that different reactions proceed on separate active patches which can be selectively poisoned is scarcely tenable. Although some reactions proceed more rapidly on certain preferred crystal planes or on lattice defects on those planes, selective poisoning would involve preferential adsorption on those planes. Since the condensation coefficient of poisons such as oxygen or sulphur is high, surface migration of the poison or of a poison–substrate complex, e.g. Ni_2O or Ni_2S would be necessary. Over the temperature range usual in catalytic hydrogenation at least such mobility does not occur. Although redistribution of poisons by desorption and readsorption is possible for weakly adsorbed poisons, its rate must necessarily be slow compared to that of the reaction itself. Again there are reactions, e.g. the ortho–para H_2 or H_2–D_2 reactions, which are not so restricted and can take place on all or on a large proportion of the catalyst surface as measured by the B.E.T. method. In a classical series of experiments Maxted showed that, in the hydrogenation of crotonic acid by platinum black in the presence of a variety of poisons, the activity of the catalyst fell first in a linear manner on progressive adsorption of the poison as determined by its distribution between liquid and surface phase; this was followed by a sharp break, resulting in a "flexed linear type" of curve. He attributed this to the presence of two types of site, all sites being equivalent among themselves in respect to catalytic activity and susceptivity to poison. These observations are not readily interpreted on the active patch hypothesis. On the other hand they follow logically according to Herington and Rideal if uniform crystal facets are regarded as the sites of catalytic activity.

Many hydrogenating metallic catalysts crystallize in the face-centred cubic form, and the reaction velocity is usually independent of the concentration of the material undergoing hydrogenation. In certain

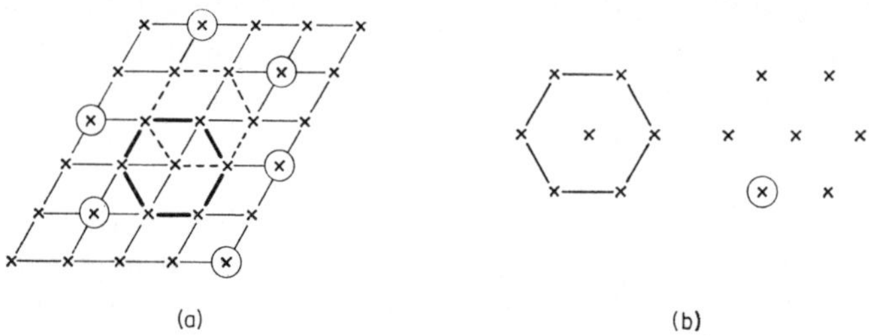

FIG. 4. (a) Interference between two possible modes of adsorption for a six point adsorbate. (b) Effect of a one point poison on a six point adsorbate. ⊗, Poisoned sites; ×, lattice sites.

cases the reaction may be of first order and some modifications in the theory are required.

The fundamental postulate is that reactant and poison are held to the surface at uniform atomic sites, the number of which occupied per molecule or atom being dependent on the composition and configuration of reactant and poison respectively.

We consider the case (Fig. 4(a)) where the reactant occupies six vacant sites, e.g. a hexagonal ring, and the poison but one site. In Fig. 4(b) the adsorbing sites (one central and six peripheral) are assumed to be so widely separated that no mutual interference is possible and there is thus a direct proportionality between the activity and the number of groups of active sites.

If, after poisoning, a fraction θ of the active sites is bare and an isolated group of n centres is required for the reactant, then the number of unpoisoned reactant-accepting groups will be proportional to θ^n. Thus for $\theta = \frac{1}{2}$ ($= 50\%$ poisoned) and $n = 2$ or 7 the activities will not be 0·5 but only 0·25 and 8×10^{-3} respectively unpoisoned. The plot of activity against poison concentration is only linear throughout the whole poison

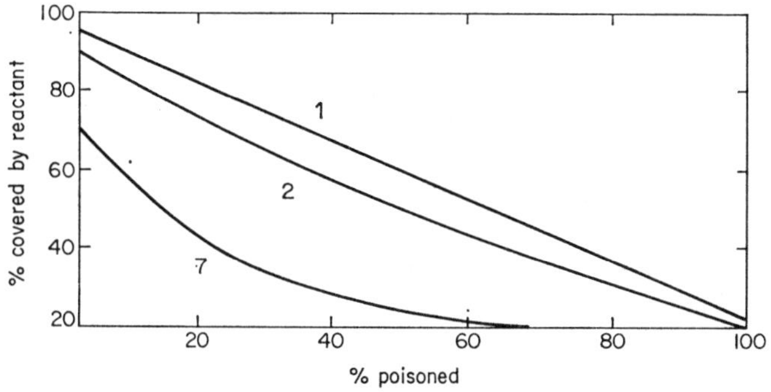

FIG. 5.

range where $n=1$. The seven point reactant has been chosen because this mode of adsorption would correspond to that of a benzene ring planar to the surface.

In Fig. 5 are shown the results of calculation for the effect of a single-point poison on the adsorption on a [111] plane of 1, 2 and 7 site reactants. The total number of sites occupied by the reactant molecules (for 1, 2 and 7 sites per molecule) is thus proportional to the number of molecules which would react in a given time for a zero order reaction. Only when reactant and poison each occupy a single site is the relationship linear, and when the poisoning becomes 100 % the activity is zero. The curve for seven-point contact reactant is very nearly linear in the poison range 0–33 % poison but thereafter shows a rather sudden curvature and at 70 % poison the ordinate is very small.

This curve could easily be interpreted as a flexed linear type. We note that at 60 % poison the catalyst is producing only 1 % of its initial conversion for a reactant with seven-point contact, whereas the activities are 31 % for two-point and 40 % for one-point contact reactants respectively. A recent example of selective poisoning is to be found in the use of sulphur compounds in the conversion of isoprene to methylbutenes and isopentane and on the ratio of the latter products to one another on a nickel catalyst. On adsorption of thiophene in the presence of hydrogen, butane is formed together with an attached sulphide ion.

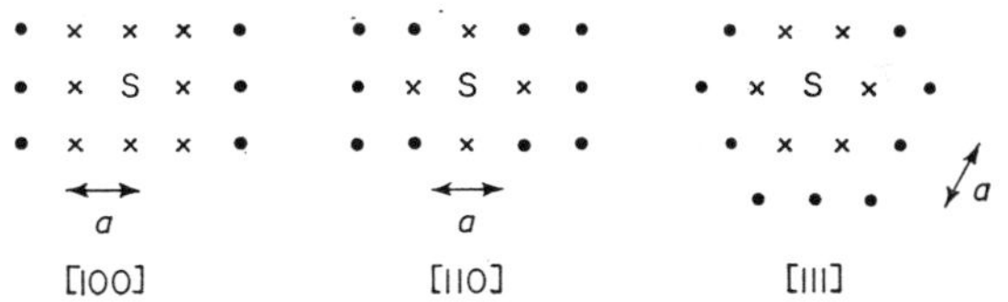

FIG. 6.

With an area of 33 Å^2 for the thiophene molecule and 1·84 Å diameter for the sulphide ion we note that, if the atomic spacings of nickel on the [100], [110], [111] planes (Fig. 6) are either 2·48 Å or 3·5 Å, the neighbours next to the sulphided site will not be able to accommodate further thiophene adsorption, and thus they are protected from poisoning. There will consequently be 8 protected atoms on the [100], 4 on the [110] and 6 on [111] planes, and there will be a maximum sulphur coverage of 10 % on the [100] plane, 38 % on the [110] and 24 % on the [111]. Further poisoning is possible with H_2S or mercaptans of cross-section smaller than that of thiophene.

Chapter 3

Condensation and Evaporation

A molecule impinging on a surface from the gas phase could be reflected, condense, condense and subsequently undergo a chemical reaction or re-evaporate after a finite life time on the surface. The rate of impact of a molecular beam of gas at pressure P is given by the Hertz-Knudsen equation:

$$\mu = \frac{P}{(2\pi mkT)^{1/2}}$$

where m is the molecular mass of the gas, and k the Boltzmann constant. If the gas is at a temperature T_1 and the surface, e.g. a wire, at a temperature T_2 then on exposure to the gas the wire will lose or gain heat. If the gas attains thermal equilibrium with the wire on impact it will leave it at the temperature of the wire. If the gas returns to the gas phase at a temperature T_2' and not T_2 we can define an accommodation coefficient

$$\alpha = \frac{T_2'-T_1}{T_2-T_1}$$

For complete thermal equilibrium $\alpha = 1$. The heat loss or gain of the wire can be measured electrically and for a monatomic gas at a pressure of P dynes cm^{-2} we find

$$\Delta H\ (\text{kcal min}^{-1}\ \text{cm}^{-2}) = \frac{1{\cdot}74\times 10^{-4}\alpha P(T_2-T_1)}{(mT_1)^{1/2}}$$

We can thus determine α from measurable quantities.

The method was first used by Eucken to determine the specific heats of gases and was used by Roberts for examining chemisorbed monolayers. Langmuir in his earlier work believed that the accommodation coefficient was in fact unity and that, owing to the "free valency" nature of the surface, all impinging molecules condensed. The molecules of physically adsorbed gases could subsequently evaporate and thus have a finite life time or *Verweilzeit* on the surface. He recorded the life time τ as the time in seconds required for one gram molecule to evaporate from one square centimetre of covered surface. Some typical values are given in Tables X and XI. If we have a mixture of gases,

TABLE X

Values of τ for various gases

	Surface				
	Liquid		Mica	Glass	
Gas	$T=90°K$	$T=150°K$	$T=90°K$	$T=150°K$	$T=90°K$
N_2	0·3	0·013	$4{\cdot}4\times10^5$	—	$2{\cdot}3\times10^5$
CH_4	5·5	0·070	$2{\cdot}4\times10^4$	—	$1{\cdot}0\times10^4$
CO	0·50	0·021	$1{\cdot}5\times10^4$	$4{\cdot}8\times10^5$	$1{\cdot}9\times10^5$
Ar	0·96	0·022	$9{\cdot}2\times10^4$	—	$7{\cdot}3\times10^4$
O_2	1·24	0·029	$8{\cdot}0\times10^4$	$9{\cdot}0\times10^3$	$7{\cdot}8\times10^4$

TABLE XI

Values of τ for carbon dioxide on platinum

Temperature (°K)	τ (sec)	Temperature (°K)	τ (sec)
300	$1{\cdot}2\times10^{11}$	650	$3{\cdot}8\times10^{-2}$
350	$5{\cdot}8\times10^{7}$	700	$6{\cdot}6\times10^{-3}$
400	$1{\cdot}9\times10^{5}$	750	$1{\cdot}4\times10^{-3}$
450	$2{\cdot}2\times10^{3}$	800	$3{\cdot}8\times10^{-4}$
500	63	900	$4{\cdot}0\times10^{-5}$
550	3·3	1000	$1{\cdot}0\times10^{-5}$
600	$3{\cdot}0\times10^{-1}$		

determination of their adsorption isotherms permits us to estimate the relative lifetimes of the molecules on the surface. This assumes that equilibrium conditions obtain, viz.

$$b_1 P_1 = \nu_1 \theta_1$$

$$b_2 P_2 = \nu_2 \theta_2$$

or

$$\frac{\nu_1}{\nu_2} = \frac{\tau_2}{\tau_1} = \frac{b_1 P_1 \theta_2}{b_2 P_2 \theta_1}$$

In some cases the presence of a second gas may augment the adsorption of the first one. We have already noted that there exist two types of adsorption, one in which the dispersive forces alone are operative, and the other in which electronic interaction is involved. In the simplest case, where the interaction is of the Morse potential type, the impinging molecule has a kinetic energy vector perpendicular to the surface, so there must be some mechanism for at least some of this energy to be lost as well as any latent heat involved in the condensation. The transfer of energy to metallic lattices is probably governed by the size of the quantum permissible by the Debye frequency, unless electronic interaction is involved. Clearly, the greater the depth of the potential trough and the slower the velocity of approach the greater will be the probability of condensation. Lennard-Jones and his co-workers have explored in detail the conditions of coupling for such energy transfer. Some values which he obtained for helium atoms striking a surface of a metal with a Debye temperature of 510°K are given in Table XII.

TABLE XII

Incident energy (cal mole^{-1})	Sticking probability × 10^2 at lattice temperature	
	30°K	100°K
1·0	7·4	9·7
16	9·0	11·7
63	9·4	12·3
142	10·0	12·6
253	10·0	12·0
569	6·5	6·9
963	0·3	0·3

The assumption that the ease of the transfer of rotational and vibrational energies for multi-atomic gases is identical with that for translation is not always justified.

The accommodation coefficient is thus not necessarily unity, and the concept of a "condensation coefficient" can be introduced to define the fraction of molecules which overcome the potential barrier.

In considering processes of condensation, great differences are to be expected when the condensed phase is mobile or immobile. Thus in the chemisorption of hydrogen on tungsten, two atoms are formed from the impinging molecule. If mobile, the adatoms will travel over the surface and separate until the repulsive forces just balance the thermal agitation. Thus at $\theta = 0{\cdot}5$ the state of the surface can best be represented by a checkerboard effect.

It is clear that subsequent filling up to $\theta = 1{\cdot}0$ will be a difficult matter, since two juxtapositioned vacancies are required to accommodate the molecule. We should also anticipate a marked change in the heat of adsorption.

The calculated comparative rates of filling for a fully mobile and an immobile film of a diatomic dissociating gas are given in Table XIII. Although the lattice can be filled completely with atoms from a mobile dissociating diatomic gas, with an immobile film we are left with ca 8% of single site isolated vacancies.

TABLE XIII

θ	$\dfrac{\text{Time of filling for immobile film}}{\text{Time of filling for mobile film}}$
0·1	1·0
0·3	1·06
0·4	1·20
0·5	2·96
0·6	$2{\cdot}8 \times 10^2$
0·7	$9{\cdot}7 \times 10^2$
0·8	$1{\cdot}9 \times 10^3$
0·9	$12{\cdot}0 \times 10^3$

For monoatomic gases such as helium, Roberts found, in agreement with the concept of energy interchange by quanta, that α varied with the temperature. For helium striking a tunsten wire he found $\alpha = 0{\cdot}025$, 0·045 and 0·10 at temperatures of 100, 200 and 300°K, respectively. More recent work has indicated that the temperature variation is in reality much smaller, of the order of 0·015–0·017 between 80° and 300°K, and that the rise in the values of α is in part due to the diffusion of impurities to the surface. If the metal surface is already covered with an adsorbed gas the condensation coefficient rises as energy transfer is facilitated; thus $\alpha = 0{\cdot}3$ for helium, and $\alpha = 0{\cdot}60$ for neon on tungsten covered with an adsorbed monolayer of hydrogen, a fact made use of by Roberts in his measurements of the surface coverage by that gas.

The condensation process is followed by chemical reaction or by evaporation after a finite lifetime. In the case of chemisorbed gases at relatively low temperatures, the lifetimes are relatively long and the "condensation coefficient" can be identified with a "sticking coefficient" as the fraction of molecules retained by the surface or at least retained during the time period of the experiment. In spite of this uncertainty, in recent years the determination of sticking coefficients or *Haftzahl* has been actively pursued either by direct measurement of the adsorption from a molecular beam of known flux or by flash desorption.

Condensation experiments with beams of metallic vapours have been carried out by a number of investigators, notably Knudsen, Wood,

Langmuir, Frenkel, Estermann, Chareton and Semenoff. These investigators found that condensation of metallic vapours did not take place on any target unless this was maintained below a certain critical temperature that was dependent on the stream density. Table XIV indicates the wide variations that have been found.

TABLE XIV

Vapour	Substrate	ν_c (atoms cm^{-2} sec^{-1})	T (°K)
Cd	Picein	2×10^{17}	203
Cd	Picein	1×10^{17}	198
Ag	Glass	3×10^{10}	192
Au	Glass	$<10^{11}$	50

The critical temperatures found for the condensation of a number of metals at identical stream densities are given approximately in Table XV. The still most satisfactory explanation for the existence of a critical stream density is that given over fifty years ago by Frenkel.

TABLE XV

Metal	Hg	Cd	Zn	Mg	I_2	Cu
T_c (°C)	−190	−90	−78	−180	−60	+350

He assumed that single atoms had a lifetime or *Verweilzeit* on the surface which was comparatively short (τ_1) whilst doublets had a far longer lifetime (τ_2). Frenkel also assumed that the target areas σ_1, σ_2 were related to the field energy required to dissociate a doublet ($\sigma_2 = \sigma_1 e^{\Delta u/kT}$).

If the total number of atoms on the surface be N, the number of single atoms will be $n_1 = N(1 - N\sigma_1)$ and the number of paired atoms $n_2 = N^2\sigma_2$.

If the molecular beam density be ν, then the rate of condensation will be

$$\frac{dN}{dt} = \nu - \frac{n_1}{\tau_1} - \frac{n_2}{\tau_2}$$

or

$$\frac{dN}{dt} = \nu - \frac{N}{\tau_2} + \beta N^2$$

where

$$\beta = \sigma_1\left(\frac{1}{\tau_1} - \frac{e^{\Delta u/kT}}{\tau_2}\right)$$

At equilibrium the rates of condensation and evaporation must be equal or

$$\nu - \frac{N}{\tau_1} + \beta N^2 = 0$$

where

$$N = \frac{\dfrac{1}{\tau_1} \pm \sqrt{\left[\left(\dfrac{1}{\tau_1}\right)^2 - 4\beta\nu\right]}}{2\beta}$$

When $(1/\tau_1)^2 = 4\beta\nu$, N must be real, i.e. there is a critical value of ν or

$$\nu_c = \frac{(1/\tau_1)^2}{4\beta}$$

The mean lifetime of the adsorbed atom can be related to the lattice vibration frequency $\nu_0 = 1/\tau_0$ by an expression of the type

$$\tau_1 = \tau_0 e^{\Delta u/kT}$$

where Δu is the energy of adsorption. The approximate form of the critical stream density then becomes

$$\nu_c = \frac{1}{4\sigma\tau_0} e^{\Delta u/kT}$$

a form of expression derived by Frenkel.

The quantitive application of these expressions to metallic condensation yields widely divergent results. Thus Chareton and Semenoff for the condensation of cadmium obtained $\Delta u = 11{,}000$ cal (g atom)$^{-1}$, and with a value at $\sigma = 30$ Å^2 we obtain $\tau = 5 \times 10^{-16}$ sec, and Estermann obtained $\Delta u = 3500$ cal (g atom)$^{-1}$ and $\tau = 5 \times 10^{-5}$ sec, and Cockroft obtained $\Delta u = 5680$ cal (g atom)$^{-1}$ and 2000 cal (g atom)$^{-1}$ for the doublet and a lifetime of $\tau = 2{\cdot}6 \times 10^{-9}$ sec. By similar calculations, values of τ from 2×10^{-3} sec for cadmium on glass to $2{\cdot}6 \times 10^{-9}$ sec for cadmium on copper have been obtained.

All these heats of evaporation are relatively small compared to the latent heats of evaporation, and we must consider that physical adsorption alone is involved. Evidently the cleanliness of the receiving surface plays the most important part in these condensations since Ditchborn and Gavin and Prague were unable to observe any critical phenomena where cadmium was deposited on clean glass over the temperature range +30 to 150°C. They support the view that nuclei form where the field is strongest, viz. on the clean portion of the glass, and nuclei formed here have a longer lifetime, than single atoms.

Surface Migration (Crystal Growth)

That crystals grow and decrease in size with well-defined facets although each plane is subject to the same stream density of impinging molecules is an indication that the condensation coefficient α cannot be unity in all cases. Discussion as to the mechanism of growth was initiated by Volmer. At equilibrium the rates of evaporation and condensation must be equal. Many attempts have been made to evaluate the rates of evaporation from different crystal facets *in vacuo* of such substances as sulphur, iodine and sodium chloride. Coefficients less than unity have been obtained and those for sulphur have been found to vary from facet to facet. It has been concluded by some that only those molecules possessing a sufficient energy of activation for deposition as well as for evaporation determine the speeds of the processes; thus a value of $E = 8{\cdot}2$ kcal mole^{-1} has been derived for the condensation of sodium chloride. It seems likely for badly conducting solids that the rate of evaporation is governed by the rate of supply of heat to compensate for the withdrawal of the latent heat of evaporation, and this can be supplied only by radiation from the environment and thermal conduction through the basal plane of the crystal which, in the case of non-isotropic crystals, will vary from facet to facet.

Volmer concluded that when a molecule impinged on the surface of a crystal it might evaporate again after a finite lifetime, but that during its lifetime in the adsorbed state it could travel over the surface and if while mobile it falls into a kink or attaches itself to an edge, condensation results. Such a process of growth can give values of condensation coefficients that are apparently greater than unity and also reveal variations from facet to facet.

Volmer and Estermann showed that in the growth of a hexagonal crystal of mercury in its supersaturated vapour the rate of extension of the basal plane could be accounted for only if the impinging molecules on those planes, calculated with the aid of the Hertz Knudsen equation, travelled over the planes to supplement those impinging and condensing on the sides of the crystal. This and similar experiments suggest that the surfaces of the crystals contain a number of molecules which can be regarded as a mobile two-dimensional amorphous phase and possessing the power of surface diffusion. Evidently, the energy of activation for surface diffusion E is less than the lattice energy for evaporation, but whether the energy for surface diffusion varies from facet to facet cannot be stated until more is known about the spacing of the edges and kinks in the respective surfaces and about the lifetimes of the atoms adsorbed on the planes.

The self surface diffusion for metals can be followed by means of radioisotopes and is found to proceed more rapidly than in the bulk phase. Recently, Blakely and Mzhura followed the process of smoothing out of interferometric scratches on a bar of metal maintained at 800–1200°C. They found that the surface diffusion constant D_s was strongly dependent on the orientation of the microcrystals, being higher on the [111] and [100] planes than on the others, and from the temperature coefficient they obtained the value

$$D_s = D_0 e^{-E/kT} = 5{\cdot}3 \times 10^{-4} e^{(-0{\cdot}62 \pm 0{\cdot}08 \text{ eV})/kT} \text{ cm}^2 \text{ sec}^{-1}$$

Surface diffusion can also be examined by electron microscopic means; thus, Müller obtained a value for the surface diffusion of tungsten on its [110] plane of $E = 30$ kcal mole^{-1}, a value to be contrasted with $\lambda_{evap} = \Delta U = 138$ kcal mole^{-1}. It is possible that this metallic surface migration is determined by lattice defect migration for the experimental value of 0·62 eV or 13,000 cal mole^{-1} is comparable to 0·79 eV for copper and 0·67 eV for gold. Here the experimental approach is to determine the resistance–temperature or specific volume–temperature curve. We are necessarily confronted with the problem of the equilibrium defect distribution between the surfaces of different crystal planes and the bulk homogeneous interior and, incidentally, the rate of exchange of a surface defect not only on the surface but also within the interior.

Spreading of Alkali Metals

Early experiments confirmed that unlike substances would also spread on solid surfaces; thus, Volmer showed that benzophenone would travel on glass but we note that neither paraffin nor cetyl alcohol would so spread. Hardy found that acetic acid would travel only by vapour condensation on dry glass but would spread when the glass was wet. A great deal of information on surface migration can be obtained by examination of the behaviour of the alkali metals on other metals such as tungsten, since photoelectric or thermionic changes in the work function are readily determined. Becker in 1929 was the first to demonstrate that barium would travel from one side to the other of a tungsten strip at 900°K and it was only at 1150°K that evaporation and recondensation played a part in the process.

Langmuir and Taylor obtained by a thermionic method values of $E = 14$ kcal (g atom)$^{-1}$ and $\lambda = 64$ kcal (g atom)$^{-1}$ for the energies of activation for surface travel and for evaporation of caesium on tungsten. Bosworth and Rideal followed the migration of potassium along a tungsten strip photoelectrically. In the deposition of metal vapours the formation of doublets accounts for the existence of a critical stream density; the view that thick films are built up from such two-dimensional nuclei has been supported by the observation that metal films deposited on glass, used so frequently for investigations on catalysts, have an area which is proportional to the weight of the metal and must therefore have a porous structure. Mitchell, who examined the deposition of copper in some detail, believes that two-dimensional nuclei are formed about 80 Å apart; these grow as hexagonal islands until the surface is covered with a monatomic layer with a hexagonal network. The nuclei for the next layer are formed at the nodes of the boundaries of the network, and thus a series of networks is built up. This cannot be the general method of formation of crystals from their vapour or solution where finite vapour pressures or solubilities are concerned for forming a two-dimensional nucleus of say sodium chloride on one of the facets of a crystal of that salt involves a degree of supersaturation to compensate for the free energy of the surface of that nucleus, i.e.

$$RT \log \frac{C}{C_{\text{sat}}} \geqslant \tfrac{1}{3}\pi r^2 \gamma$$

where γ is the interfacial free energy. For any reasonable value of γ and of nucleus size the degree of supersaturation for growth must be at least 1·5 times saturation. In point of fact, however, the cubic crystals will commence to grow when the supersaturation is as low as 0·045%. Evidently, already existing dislocations and steps play an important part in the growth process.

Surface diffusion is caused by a surface spreading pressure, and since this cannot be measured directly it is convenient to define the surface pressure π in terms of the thermodynamic potential of the system, i.e.

$$\pi = \frac{\mathrm{d}G}{\mathrm{d}\mu}$$

where G is the Gibbs function. We can resolve this into kinetic π_k and potential π_p terms: thus

$$\pi = \pi_k + \pi_p$$

where the kinetic contribution is $\pi_k = n\sigma kT$, n being the fraction of activated molecules in the surface plane of concentration σ; the potential term is $\pi_p = \frac{1}{2}\sigma \sum rf$, where f is the force between any pair of individual adsorbed atoms at a separation r. If E is the activation energy for migration in a layer of surface density σ we have:

$$d\pi = -\sigma dE$$

$$\pi = -1{\cdot}58 \times 10^{-12} \int \sigma dE \text{ dynes cm}^{-1}$$

where E is measured in electron volts.

Further, from the vapour pressure of the material in equilibrium with the adsorbed film we obtain from the Gibbs adsorption isotherm:

$$\frac{d\pi}{d \ln P} = \sigma kT$$

The experiments of Langmuir with thorium and of Bosworth with sodium and potassium on tungsten show that the Fick surface diffusion constant D_s decreases with increase of σ and at the same time the activation energy associated with D_s falls, so that less work is required to cause a concentrated film to spread than a dilute one. It is interesting to compare the spreading pressure of potassium on tungsten calculated by the above equation with those say of formic acid on mercury (Table XVI). The great increase in spreading pressure with concentration for the

TABLE XVI

Surface density σ ($\times 10^{-14}$ molecules cm^{-2})	π (dynes cm^{-1}) K on W	π (dynes cm^{-1}) HCOOH on Hg	Activation energy E for K on W (kcal (g atom)$^{-1}$)
0·5	9	9	14·7
1·0	22	16	12·1
1·5	60	22	10·3
2·0	70	28	—
3·0	187	38	7·2
4·0	190	48	—
4·6	307	52	6·6

alkali metal indicates that the intermolecular repulsive forces are all important. We have seen that the adatoms of potassium on tungsten can be regarded as dipoles. Topping showed that the repulsive forces of an array of dipoles is given by:

$$\pi_p = 4{\cdot}51\mu^2 \sigma^{5/2}$$

On substitution of the dipole value for the surface potential in the Topping equation we obtain

$$\pi_p = 0{\cdot}114(\Delta V)^2 \sigma^{1/2}$$

where ΔV is measured in e.s.u. The validity of this treatment has been discussed by Langmuir. Table XVII gives the values of π calculated from surface migration experiments on the assumption that $\pi_p = \pi - \sigma kT$ and from Topping's expression using photoelectric changes in the work function to evaluate σ.

TABLE XVII

Surface density σ ($\times 10^{-14}$ atoms cm^{-2})	π_p (dynes cm^{-1}) Migration	π_p (dynes cm^{-1}) Topping
0·5	8	9
1·0	19	21
1·5	40	40
2·0	65	62

That electrostatic interaction accounts for almost all the surface free energy of alkali metal films is further confirmed by some experiments by Bosworth on the vapour pressure above sodium films with simultaneous measurement of the contact potential and making use of the Gibbs adsorption equation (Table XVIII). Some values for the surface diffusion coefficients of caesium on tungsten are given in Table XIX (Langmuir).

TABLE XVIII

Surface density σ ($\times 10^{-14}$ atoms cm^{-2})	π (dynes cm^{-1}) Gibbs	π_p (dynes cm^{-1}) Topping
0·27	1·4	3·6
0·54	5·6	13·0
1·08	22·0	42·0
2·16	90·0	110·3
2·70	140	153

TABLE XIX

Fraction covered θ	D_s (cm^2 sec$^{-1}\times 10^5$)	T (°K)	D_s at $\theta=0{\cdot}4$ (cm^2 sec$^{-1}\times 10^5$)	τ (sec $\times 10^5$)
0·15	60	300	$1{\cdot}2\times 10^{-6}$	1·5
0·057	11	400	$4{\cdot}3\times 10^{-4}$	$4{\cdot}1\times 10^{-3}$
0·037	5	500	0·015	$1{\cdot}2\times 10^{-4}$
		600	0·16	$1{\cdot}1\times 10^{-5}$
		700	0·8	$2{\cdot}1\times 10^{-6}$
		967	11	

Energy of activation for migration at $\theta_{0\cdot 03}$ of $E=0{\cdot}61$ eV,

For potassium on tungsten the data in Table XX were obtained by Bosworth and Rideal.

TABLE XX

σ ($\times 10^{-14}$ atoms cm^{-2})	E (eV)	T (°K)	D_s at $\theta = 0{\cdot}4$ (cm^2 sec^{-1} $\times 10^5$)
0·06	0·69	480	0·57
0·48	0·59	510	1·4
1·2	0·52	590	10
2·4	0·35	610	16
3·9	0·33	710	140
4·8	0·29	780	280

The changes in π produced by a change in surface concentration can be expressed by two different methods from mobility changes and from the Gibbs equation

$$\underset{\text{Mobility}}{\mathrm{d}\pi} = -\sigma \mathrm{d}E$$

$$\underset{\text{Gibbs}}{\mathrm{d}\pi} = \sigma kT \,\mathrm{d}\ln P$$

on setting these two derivatives equal we obtain, since we are dealing with an essentially isothermal process

$$-\frac{\mathrm{d}E}{\mathrm{d}\sigma} = \frac{\mathrm{d}\left(\dfrac{k\ln P}{1/T}\right)}{\mathrm{d}\sigma}$$

Furthermore, $k\ln P/(1/T)$ can, with the exception of a constant which will disappear on integration, be indentified with $-\lambda$, the heat of evaporation at a concentration σ. The equation can therefore be written:

$$\frac{\mathrm{d}E}{\mathrm{d}\sigma} = \frac{\mathrm{d}\lambda}{\mathrm{d}\sigma}$$

or

$$\lambda - E = \text{constant}$$

In Fig. 7 the level A is the lowest level that an adatom can occupy on the surface and is normally occupied by the first atom adsorbed. As the surface concentration rises, the adatoms take up higher levels so that at concentration σ the level A′ is occupied and at the highest concentration possible or the concentration of greatest π the level A″ is occupied.

The level B represents the mobile level as on activation to this level an atom is able to move freely over the surface. C is the lowest level in the gas phase and activation to C results in evaporation of the adatom. The influence on the filling up of the levels with progressive adsorption on the mobility of the adsorbate is well demonstrated in the case of oxygen on tungsten. At 90°K the process is one of immobile condensation. At 1100°K the curve is quite different and corresponds to one of mobile condensation. At an intermediate temperature, say 600°K, we find that the condensation initially proceeds along the curve for the immobile film but at a certain σ switches over to that for a mobile film. The lower the temperature at which the condensation is studied the higher the concentration at which the change takes place. It is interesting to note

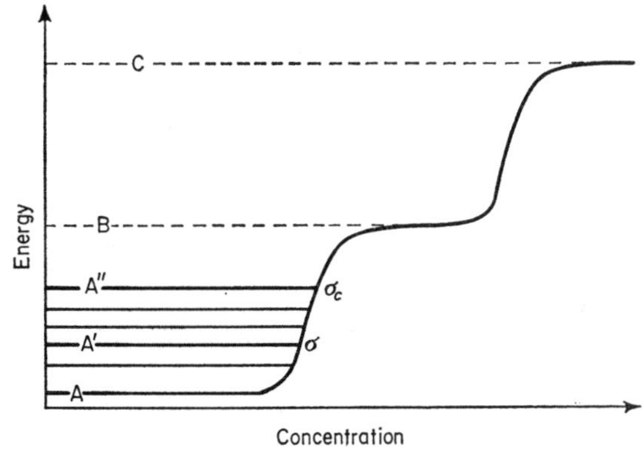

FIG. 7.

that sodium will diffuse into tungsten but potassium does not do so at all readily. This diffusion inwards probably proceeds through the defects and grain boundaries. In support of this view it is found that, if a condensed (two-dimensional) liquid film be evaporated, the resulting vapour film possesses a much greater surface mobility than does one produced by condensation of an equal amount of vapour, suggesting that the film left on partial evaporation is not uniform but consists of a liquid condensed network along the grain boundaries and cracks together with a much less sparsely covered surface.

If the equilibrium vapour pressures be known for a definite coverage θ and the condensation coefficient α is taken as unity, the mean life of the condensed atom on the surface will be given by

$$\frac{n\theta}{\tau} = \frac{P}{(2\pi mkT)^{1/2}}$$

The adsorbed atoms may be regarded as evaporating independantly as from a two-dimensional gas in accordance with a simple unimolecular law.

$$-\frac{dn}{dt} = n\theta\left(\frac{kT}{h}\right)\frac{1}{f}e^{-\lambda/kT}$$

Where f is the vibrational contribution to the partition function. Ehrlich (1954) examined the validity of these assumptions by calculation of the constant kT/h from experimental data. Some of his values are given in Table XXI for adsorption on tungsten. The lifetimes of the species are thus of the order of the vibrational frequency of the underlying lattice ν.

TABLE XXI

Substance	λ (kcal mole^{-1})	$1/\tau$ (10^{12} sec^{-1})	kT/h (10^{12} sec^{-1})
Na	62·9	5·13	2·7–3·0
K^+	58·8	33·1	1·1–1·5 = $kT/2h$
Rb	60·0	4·0	2·9–3·3
Cs	64·4	1·75	0·95–2·1
Ba	87·2	0·038	2·5–2·8
H	67	22	2·5–3·3
O	147	12·5	4·9–5·4

Of recent years the field emission electron microscope has been utilized for the determination of the rates of surface diffusion of adsorbed atoms, notably by Müller and Gomer (1959). Some of the values obtained for adsorption on tungsten by this method are given in Table XXII.

TABLE XXII

Substance	E (kcal mole^{-1})	λ (kcal mole^{-1})
O_2	30	150
H_2	20	160
Xe	3·8	10
A	0·6	1·9
CO	30	80

The examination by field emission on deposits of Au and Cu on tungsten reveals that they form epitaxial layers. For Au the first monolayer is accompanied by a rise in the electron work function from $\phi = 4{\cdot}52$ eV to 5·12 eV i.e. the dipole is Au^-W^+. This can spread with an activation energy of 0·70 eV and $\lambda = 4{\cdot}06$ eV. On this monolayer a second layer can be built up with a decrease in ϕ to 4·92 eV. Gold in excess of this can be deposited in epitactic form up to at least 20 layers. This possesses the work function of gold and its λ is also that of gold. Crystallites of gold were found to be metastable with respect to this epitactic form. It is interesting to note that the ratio E/λ is roughly 0·25 for a number of these surface diffusion processes.

It was shown by Kozeny that the rate of flow of a fluid through a packed powder was given by the equation

$$\frac{\mathrm{d}n}{\mathrm{d}t} = \frac{pa}{k\eta l}\frac{V^3}{(1-V)^2}\frac{1}{A^2}$$

where A is the area of the particles per cm^3 of solid, V is the void fraction in the column, p is the pressure difference, η the viscosity, l is the equivalent capillary length and a is the area of the column. Subsequent investigations, notably by Carman and by Barrer (1966), have shown that the Kozeny equation requires modification to fit experimental results, especially for readily adsorbed vapours. The equation rests on the premise that the Poiseuille equation is applicable to the flow down a single capillary; this assumes that the contour of flow velocity is parabolic, being zero at the walls. If, however, there is a slip at the walls or, in other words, a two-dimensional or surface migration, the surface transport may well become commensurate with the bulk transport when the capillaries are small. From such measurements Carman found a value of $D_s = 5 \times 10^{-5}\ \mathrm{cm}^2\ \mathrm{sec}^{-1}$ for n-butane passing through porous silica gel at $-10°\mathrm{C}$. The surface flow is regarded by some as a hydrodynamic or viscous one, by others as a diffusive or random walk process in which motion is the result of hopping. The experimental evidence on the whole favours the latter assumption for chemisorbed systems.

In all probability the hydrodynamic flow is transformed at low pressures to independent molecular movement as in a two-dimensional gas, where the mean free path of an adsorbed molecule is governed by intermolecular collisions and not by hopping as in chemisorbed systems. Pope has recently shown that SO_2 diffuses over spheron in this manner.

On the Mechanism of Motion

The assumption that physical or chemisorbed species are either fixed on sites or behave as a free two-dimensional gas the conversion of one form to the other being dependent on the temperature and surface concentration is in all probability an oversimplification of the case. Certainly in the case of the alkali metals processes of surface diffusion are involved and we may formulate several mechanisms for this diffusive process. The adatom may move into the next contiguous vacant site or it may be raised to a migratory level and travel some distance before de-activation, either by loss of energy or by collision before falling back to a fixed site after traversing a mean free path. This path may be independent of the direction or, as we suspect from earlier work on self diffusion in metals, it may have preferred paths dependent on the crystal facet over which it is travelling.

The kinetic two-dimensional diffusion constant can be written

$$D_s = \tfrac{1}{2}\lambda v$$

where λ is the mean free path and v is the velocity of the diffusing particle. If, as we have seen, the diffusion can be regarded as an activated process, the probability of a particle being in the mobile state will be $e^{-E/kT}$. Thus the diffusion constant averaged over all the particles, mobile and static, will be $D_s = \frac{1}{2}\lambda v\, e^{-E/kT}$. If we assign the kinetic value to v, i.e. $v = (3RT/m)^{1/2}$ where m is the molecular mass

$$D_s = (3R/4)^{1/2}\lambda(T/m)^{1/2} e^{-E/kT} = 7{\cdot}87 \times 10^2 T^{1/2} m^{-1/2} e^{-E/kT}$$

For sodium on tungsten $D_s = 8 \times 10^{-6}$ cm^2 sec^{-1}, $m = 23$, $T = 293$°K, $E = 2900$ cal mole^{-1}, and we obtain $\lambda = 5{\cdot}7 \times 10^{-6}$ cm, or the traverse in the free state is some hundred atoms per hop. This assumes that in flight the particle has the mean gas kinetic velocity, which is debatable. The energy of activation E may be temperature dependent or

$$D_s = D_{so} T^{(\frac{1}{2}+y)} e^{-E_0/kT} \tag{1}$$

and values of y for specific mechanisms of diffusion may be obtained with the aid of the Kirchhoff equation

$$-y = -\frac{dG}{dT} = \frac{1}{R}(C_p - C_p')$$

where C_p and C_p' are the specific heats in the static and mobile states, respectively. If the motion perpendicular to the surface in the two states is the same and can be neglected for a rigid adsorbed particle, $C_p = 2k$ and, for a freely moving particle, $C_p' = k$. If constrained to move along a line of equipotential $C_p' = 3k/2$. In this latter case we find

$$D_s = 2D_{so} e^{-E_0/kT} \tag{2}$$

and for a freely moving particle

$$D_s = D_{so} T^{-1/2} e^{-E_0/kT} \tag{3}$$

Experimental data are not sufficiently numerous or exact to permit a decision between (1), (2) and (3), but it is worth noting that constrained migration equation (2) appears to give the best agreement. It is frequently assumed that the particle can be represented as having a lifetime τ_a in the adsorbed state and τ_m in the migratory state. The diffusion constant would then be given by $D_s = \frac{1}{2}v^2[\tau_m^2/(\tau_m + \tau_a)]$. At low temperatures $\tau_a \gg \tau_m$ and we might expect the mean free path to be small and of the order of a lattice spacing, but at high temperatures $\tau_m \gg \tau_a$ and all atoms migrate. If collision results in deactivation, we obtain $D = \frac{1}{2}v\lambda$.

Chapter 4

Thermal Changes and Reaction Kinetics

Chemi-adsorption Isotherms

The basic assumptions involved in the Langmuir adsorption isotherm, namely, independence of the processes of condensation and of evaporation and independence of the rates of both processes on the extent of surface coverage, are fulfilled in an unexpectedly large number of cases. On the assumption that in the adsorption of a molecule or atom only one "elementary square" or "site" on the adsorbing surface is involved, we may write for the rates of condensation and of evaporation

$$\omega_{\text{condensation}} = \alpha(1-\theta)\,\mu$$

$$\omega_{\text{evaporation}} = \nu\theta$$

where α is the condensation coefficient, θ the fraction of the surface covered and ν is the specific rate of evaporation. In the steady state

$$\omega = \nu\theta = \alpha(1-\theta)\,\mu$$

or

$$\theta = \frac{\alpha\mu}{\nu+\alpha\mu} = \frac{bP}{1+bP}$$

where b involves the two Langmuir constants α and ν for the system. The value of μ can be obtained from the Hertz-Knudsen equation

$$\mu = \frac{P}{(2\pi mkT)^{1/2}}$$

Some consideration must be given to the concept of "elementary square" or "site". On crystalline surfaces the number of metal atoms in the surface plane will evidently depend on the plane exposed. Further, in body-centred cubic systems there will be exposed an array of atoms below the surface plane.

We cannot necessarily identify a "site" with a "bond" between an atom in or on the surface and the adsorbate but rather with the position of minimum free energy which, on a cubic crystal surface, may well be in the depression at the centre of four atoms in the crystal lattice surface plane. Complete coverage at $\theta=1$ will thus correspond to the number of surface atoms of metal exposed. Evidently, the monolayer capacity will vary with the plane exposed. Brennan has made a close study of the monolayer capacities of metals for the gases krypton and xenon at low temperatures where the gases are on fixed sites. Geometrically, we obtain the pattern shown in Table XXIII. These capacities assume that there is no interaction between the adatoms, but in general there will be some

TABLE XXIII

Plane	Monolayer capacity in terms of the interatomic distance a	
	Face-centred cubic	Body-centred cubic
[100]	$1/a^2$	$1/2a^2$
[110]	$1/\sqrt{2}a^2$	$1/\sqrt{2}a^2$
[111]	$4/3\sqrt{3}a^2$	$\sqrt{2}/3a^2$

lateral interaction which will be dependent both on the lattice spacing and on the adsorbate. The lateral interaction between the adsorbate molecules was calculated by Brennan using a variety of potential functions of which that due to Lennard-Jones ($E=(a/d^9)-(b/d^6)$) is most generally employed. Repulsive forces between 1,2 neighbours which may be diminished by attraction between 1,3 and 1,4 near neighbours were found to be great enough, for these gases on metals in which the atomic separations were small, to prohibit the formation of complete monolayers in certain configurations. In contrast to physical adsorption the heats of chemisorption are, however, usually great enough to outweigh quite large interadsorbate repulsions and ensure monolayer completion. We may note that although the original concept of surface free valencies results, on completion of the monolayer, in the formation of a saturated surface, this model presents a surface in which the surface metallic atoms are by no means completely shielded, and second layer formation is possible. In general, uniformity of the adsorbing surface only obtains on liquids. On solids the concept of complete irregularity is probably inapplicable and we are presented with surfaces which may be uniform but which contain small irregularities or active patches, lattice defects, or surfaces that consist of relatively large areas of regular but different structures corresponding to different crystal facets. These may be treated as the summation of two or three Langmuirian isotherms with the addition of a small constant to represent the strongly adsorbing active patches. Experimentally, histograms can be plotted of the departures in the P–θ curve and from a first approximate examination we can obtain the characteristic constants α and ν for the separate isotherms.

Many modifications of the Langmuir equation have been introduced as circumstances demand. On the adsorption of a non-dissociating diatomic molecule or of carbon monoxide in the bidentate form where two elementary sites per molecule are required we have as the relationship between adsorption and desorption rates

$$\alpha\mu(1-\theta)^2 = \nu\theta^2$$

whence

$$\theta = \frac{\left(\frac{\alpha\mu}{\nu}\right)^{1/2}}{1+\left(\frac{\alpha\mu}{\nu}\right)^{1/2}}$$

or at low pressures the coverage is proportional to the square root of the pressure. We note also that completeness of coverage is possible only if the adsorbed molecules are mobile, since isolated sites cannot be filled by a diatomic adsorbate, but can by the subsequent addition of a monatomic gas. Little work has been carried out on the composition of the surface phase resulting from the adsorption of mixtures of gases. In the simplest case, where each gas undergoes a simple Langmuirian adsorption independent of the other, we can write

$$\omega_1 = \alpha_1\mu_1(1-\theta_1-\theta_2) = \nu_1\theta_1$$

$$\omega_2 = \alpha_2\mu_2(1-\theta_1-\theta_2) = \nu_2\theta_2$$

whence

$$\theta_1 = \frac{\alpha_1\mu_1}{1+\frac{\alpha_1\mu_1}{\nu_1}+\frac{\alpha_2\mu_2}{\nu_2}}$$

and

$$\theta_2 = \frac{\alpha_2\mu_2}{1+\frac{\alpha_1\mu_1}{\nu_1}+\frac{\alpha_2\mu_2}{\nu_2}}$$

where subscripts 1 and 2 refer to the two gases. These respective fractions of the surface covered may be greatly modified if there are any lateral interactions. Dependent on the magnitudes of E_{12}, E_{11} and E_{22}, the lateral interaction energies, the adsorption of one gas may increase or decrease the relative adsorption of the other. Again the second layer formation may be possible under the P, T conditions of the experiment for one gas and not for the other.

We have already noted that this problem of surface availability is the most important factor operative in the poisoning of catalysts. In many cases there is lateral interaction between the adsorbed molecules, and the appropriate isotherm as a function of the number of nearest neighbours Z can be written

$$P = K\frac{\theta}{1-\theta}\left(\frac{2(1-\theta)}{\beta+1-Z\theta}\right)^Z$$

Fowler examined the modifications required if both short and long range forces between the adsorbate molecules were taken into account. Writing the free energies in the form

$$F = F_0 - fZ$$

the long range forces being considered to include both polarization and dipole repulsion, he obtained the relationship

$$\frac{\theta}{1-\theta} = P\frac{h^3}{(2\pi m)^{3/2}(kT)^{5/2}}\frac{f_a}{f_g}e^{F/kT}$$

where f_a and f_g are the partition functions of the adsorbed and free gas and F is related to ΔH the overall heat of adsorption by the equation

$$\Delta H = F - U_g - U_a - f(Z)$$

where U_a and U_g are the mean free energies per gram mole of gas in the free and adsorbed states, and $f(Z)$ is the interaction energy. If there is equipartition of energy between rotation and translation U_g and U_a will be small, and when there is no interaction $f(Z) = 0$. Hence

$$F = \Delta H - RT$$

Kiselev has derived an adsorption isotherm for localized monolayer adsorption in terms of the equilibrium constants K_1 and K_2 of the adsorbate–adsorbent and adsorbate–adsorbate interactions. This takes the form

$$\frac{P}{P_{\text{sat}}} = \frac{\theta}{K_1(1-\theta)(1+K_2\theta)}$$

If the adsorbed monolayer is regarded as mobile with one or two degrees of freedom, and there is also lateral interaction between the adsorbed molecules, the appropriate adsorption isotherm becomes

$$P = K\frac{\theta}{1-\theta}e^{\theta/(1-\theta)}$$

In many cases the adsorbed phase has been regarded as a two-dimensional gas obeying an equation of the van der Waals type

$$\left(\pi+\frac{\alpha}{A^2}\right)(A-A_0) = RT$$

and we obtain

$$P = K\frac{\theta}{1-\theta}\mathrm{e}^{\theta/(1-\theta)-2\alpha\theta/RTA_0}$$

We note that the critical temperature of this two-dimensional gas would be $T_c = 8\alpha/27RA_0$, which is much less than its three-dimensional value, and the two-dimensional compressibility $\frac{1}{A}\left(\frac{\partial A}{\partial P}\right)_T$ would be similar to that in bulk. In the case of chemisorption, however, where dipole formation is involved, the two-dimensional interactions are repulsive rather than attractive and, in addition, the tacit assumption that the mean two-dimensional molecular velocity is equal to that in the gas phase is certainly not justified. No exact measurements appear to be available for the adsorption isotherms of a gas over a range of temperature sufficiently wide to cover the transition from a fixed to a mobile phase, and the problem whether there is a critical temperature of transition or a Boltzmann distribution of free and bound particles over an extended temperature range as yet cannot be answered. In the Langmuir isotherm a change in temperature will be reflected in changes both in the condensation coefficient α as well as in the lifetime $\tau(1/\nu)$ of the adsorbed species. If we regard these changes as due to the intrusion of energies of activation both for adsorption and desorption we obtain for the process of condensation

$$\omega = \frac{P(1-\theta)}{(2\pi mkT)^{1/2}}\alpha_0\,\mathrm{e}^{-E_{\mathrm{ad}}/kT}$$

and for evaporation

$$\omega = \theta\nu_0\,\mathrm{e}^{-E_{\mathrm{de}}/kT}$$

The heat of adsorption $\Delta H = E_{\mathrm{ad}} - E_{\mathrm{de}}$ can thus be obtained. In many cases the energies of activation for condensation and for evaporation are not independent of the coverage. Thus the process of chemisorption may be retarded as the surface coverage increases, because of the decrease in free energy of the surface. If expressed as equivalent to a rise in energy of activation for condensation we obtain

$$E_{\mathrm{ad}} = E_{0,\mathrm{ad}} + \alpha\theta$$

If this additional barrier is to be attributed to the formation of an electric double layer E_{ad} would take the form

$$E_{\text{ad}} = E_{0,\text{ad}} + \alpha\theta^{3/2}\mu^2$$

resulting in a nonlinear ΔH–θ curve. On evaporation one may anticipate a concomitant fall in E_{de} with coverage or

$$E_{\text{de}} = E_{0,\text{de}} - \beta'\theta$$

and

$$\Delta H_0 = E_{\text{ad}} - E_{\text{de}} + (\alpha - \beta')\theta$$

$$= E_{\text{ad}} - E_{\text{de}} + \gamma\theta$$

The isotherm can then be written

$$\frac{\theta}{1-\theta} = \frac{\alpha_0}{\nu_0} e^{(E_{\text{ad}} - E_{\text{de}})/kT} \frac{P}{(2\pi mkT)^{1/2}}$$

$$= \frac{b_0}{(2\pi mkT)^{1/2}} e^{(\Delta H_0 + \gamma\theta)/kT} P$$

From which, by taking logarithms, we obtain

$$\log\left(\frac{\theta}{1-\theta}\right) = \log\left(\frac{b_0}{(2\pi mkT)^{1/2}}\right) + \frac{\Delta H_0 + \gamma\theta}{kT} + \log P = K + \frac{\Delta H_0 + \gamma\theta}{kT} + \log P$$

In the neighbourhood of $\theta = 0{\cdot}5$, $\log(\theta/(1-\theta))$ can be neglected, and we obtain a linear relationship between $\log P$ and θ which has been used by numerous investigators, notably Roginsky, Zeldovitch, Elovitch and Temkin, both in respect of the velocities of adsorption and desorption as well as of the P–θ isotherms themselves. The rate of dissociative adsorption of a diatomic gas involving a linear dependence on the energy of activation would thus be given in the Elovitch form by

$$\omega = \frac{\alpha P}{(2\pi mkT)^{1/2}} (1-\theta)^2 e^{-(E_0 + \alpha\theta)/kT}$$

On Heats of Adsorption

The heat of adsorption is in general greater for chemisorption than for physical adsorption, the latter being greater than but approximating to the heat of liquefaction. There are three methods by which the heats of adsorption may be determined, the most direct being calorimetric. Here both adiabatic and isothermal or phase-transition calorimeters as well as microcalorimeters have been employed. In some cases it is possible to use the adsorbent as its own calorimeter, such as measurement of the change in temperature of a wire on admission of the gas. The heat liberated when small doses of gas are admitted together with the data for the amount of gas adsorbed at the final pressure attained by the gas admitted permit one to determine the ΔH_c–θ curve where ΔH_c is the calorimetric differential heat.

When the P–θ curves have been determined over a range of temperatures the isosteric heat of adsorption ΔH_i can be determined with the aid of the Clausius-Clapeyron equation

$$\Delta H_i = RT^2\left(\frac{\partial \ln P}{\partial T}\right)_\theta$$

The maximum external work done on the gas during adsorption is RT and, in consequence, the true differential heat of adsorption ΔH_d is obtained from ΔH_i by the relationship

$$\Delta H_d = \Delta H_i - RT$$

The calorimetric heat of adsorption ΔH_c lies between the values of ΔH_d and ΔH_i, the fraction of the external work expended in the calorimeter being dependent on the design and technique of operation of the calorimeter. The change in enthalpy with temperature $(\Delta H/\Delta T)$ $(\Delta C_p/\Delta T)$ can be determined by either method. The heat of adsorption may also be evaluated by kinetic methods in which the velocities of adsorption and of desorption are determined over a range of temperature. This method has been used with success for evaluating the ΔH–θ curve for the adsorption of hydrogen on carbon. At ordinary temperatures hydrogen is reversibly adsorbed by carbon with a value $\Delta H = \sim 900$ cal mole^{-1}, the heat being practically independent of the surface coverage. At higher temperatures, however, there is again a rapid reversible adsorption of hydrogen. This adsorption involves activation energies both for adsorption and desorption. These were found to be

$$E_{\text{ad}} = E_{0,\,\text{ad}} + \alpha\theta = 79 + 14\theta \text{ kcal mole}^{-1}$$

$$E_{\text{de}} = E_{0,\,\text{de}} - \beta\theta = 137 - 42\theta \text{ kcal mole}^{-1}$$

The heat of adsorption of hydrogen is consequently

$$\Delta H_\theta = E_{\text{de}} - E_{\text{ad}} = 58 - 56\theta \text{ kcal mole}^{-1}$$

This heat of adsorption must be attributed to a chemisorption process, the entity formed being in all probability a C—H linkage since further hydrogenation results in the formation of methane. The fall in the heat of adsorption as $\theta \to 1$ is by no means an uncommon phenomenon in chemisorption processes and is open to various interpretations. In this particular case of hydrogen on carbon, the view that it is an example of localized adsorption with lateral interaction has been advanced as well as the alternative that it is a mobile chemisorbed system, the surface being non-uniform in character.

Similar difficulties are met with in considering the data for the heats of chemisorption of gases on metals. Recent work on the adsorption of hydrogen on tungsten suggests that $\Delta H = 35$ kcal mole^{-1} from θ_0 to θ_1 at 90°K but that above 195°K $\Delta H = 46$ kcal mole^{-1} at θ_0, falling to 35 kcal mole^{-1} at $\theta_{0\cdot5}$ and to 22 kcal mole^{-1} at $\theta_{0\cdot8}$.

On nickel, similarly, the value of $\Delta H = 22$ kcal mole^{-1} at 90°K but falls from 25·6 to 19 kcal mole^{-1} at 273°K, according to some investigators, but remains constant at $\sim$19·5 kcal mole^{-1} according to others. We have already noted that on adsorption of a gas a double layer comes into existence and the observed drop in the heat of adsorption with coverage has been attributed by de Boer to this increase in the electrical energy of the system. He found that the heat of adsorption on his model was given by the expression

$$\Delta H = 25\cdot1 - 14\cdot4\theta \text{ kcal mole}^{-1}$$

and that the change in work function from θ_0 to θ_1 was $-0\cdot345$ volt. On this view the double layer contribution to the energy is

$$e\Delta\phi = 7\cdot95 \text{ kcal mole}^{-1}$$

and the heat of adsorption per gram atom of hydrogen on nickel becomes

$$\Delta H_\theta = \tfrac{1}{2}(D_{H_2} + 25\cdot1 - 14\cdot4\theta) = 64 - 7\cdot2\theta \text{ kcal mole}^{-1}$$

The difficulty in ascribing the building of the double layer to a process separate from that of building the bond is that these changes in ΔH with θ should not be so temperature dependent as they appear to be. If the surface is non-uniform, the heat of adsorption is evidently dependent on the mobility of the adsorbed phase. At low temperatures when the surface phase is immobile, coverage of the whole area will proceed at a constant rate and the heat of adsorption should remain constant with respect to θ. Two different aspects of this process of irreversible condensation on a non-uniform surface may be advanced. Either that all condensations are irreversible or that there is surface mobility and condensation proceeds at the condensed phase–metal interface, originating in condensation centres on the substrate. We have had occasion to note that copper monolayers appear to grow in this manner. On elevation of the temperature, if mobility commences not only will there be a migration from the less active to the more active portions of the surface, resulting in a higher heat of adsorption at the commencement of the adsorption process, but there will also be a diminution in the entropy change as the adatoms become mobile which will manifest itself in an apparent fall in the heat of adsorption. These changes from a constant to a falling heat of adsorption on elevation of the temperature have been reported for the adsorption of hydrogen both on tungsten and on nickel. Although changes in the heat of adsorption with coverage receive ready interpretations on the assumptions that there is surface interaction, or that non-uniform surfaces are present on the catalyst, with mobile adsorption we note that there is another factor, namely bulk solution, that may give rise to a decreasing heat of "sorption" on elevation of the temperature. The hydrogenating metallic catalysts can readily "dissolve" hydrogen and the rate of solution has been determined for some of the more important ones. In some cases, e.g. in palladium, the state of the "dissolved" hydrogen in the metal has been established whereas with nickel and iron the molecular hydrogen is certainly dissociated in the metal and there are appreciable changes in the conductivity. Solution proceeds at rates that increase exponentially with temperature and when the experimental conditions for the determination of the calorimetric or P–θ data are such that solution is taking place evidently errors are introduced both in respect of the amount of gas adsorbed, of the difference between the heat of adsorption and heat of solution and of the different electronic states and lattice spacing between the metal and the metal containing hydrogen. Thus these all contribute to a falling heat of adsorption with elevation of the temperature. Although the chemisorption of oxygen on metals is frequently associated with a rise in work function and a fall in heat evolution with rising θ, when a separate phase can be identified (e.g. PtO_2, RhO_2, RuO, Ir_2O_3 at high temperatures and many other metals at low temperatures), the heat of formation is independent of the quantity of oxygen consumed, suggesting growth being confined to the oxide–metal boundary. Brennan has determined the calorimetric heats of adsorption of CO on a number of metals by Beeck's method and obtained the following values in kcal $mole^{-1}$ for the integral heats of adsorption: Ti, 150; Rh, 41; Pt, 44; Fe, 35; Zr, 148; Ta, 128; Pd, 40; Co, 40; Ni, 40. These were almost independent of θ. Falling heats of adsorption were found with W, Mo, Ni, Mn. Although a large heat liberation on adsorption implies a strong bonding of the adsorbed molecule or atom to the substrate and thus a large energy of activation for any subsequent reaction of the adsorbed species, the reverse case where the heat liberation is small does not necessarily imply a highly reactive adsorbed species. Thus, in the surface dissociation of hydrogen on a metal the M—H bond formation energy may be very large although the molecular heat of adsorption may be negligible or even endothermic.

On Entropy Change on Adsorption

The adsorption of a gas by a substrate results in constraints being placed on the molecules so adsorbed or a decrease in the entropy. These constraints are weaker for molecules physically adsorbed than for those chemisorbed. In considering the diminution in the degrees of freedom of a freely moving gas molecule on adsorption it is evident that at least one degree of translational movement must be lost resulting in a two-dimensional gas. At the other extreme, when chemisorption is involved and the molecules are attached on fixed sites, all three translational movements are lost. There may be a partial replacement of these by vibrational movements and the adsorbed molecule may or may not still be capable of rotation about one of its axes. Furthermore, with molecules on fixed sites there will be an entropy change due to the various ways in which the molecules can be distributed on the sites available.

Finally, as the surface coverage changes with change in pressure or temperature there may be a conversion of one type of adsorption to another, either gradual or as a phase change or, again, dissociation or association may take place even when the adsorption is localized. Some information on all these points can be obtained if the entropy change on adsorption can be evaluated. Thus, in order to evaluate the entropies a knowledge of the free and total energy changes involved on adsorption to a defined standard state is required, as well as of how the standard state is related to the surface coverage. Several proposals have been made to define an appropriate standard state. A standard reference surface pressure π_s has been used to evaluate the entropy change involved in desorption from this two-dimensional pressure to the ideal gas pressure $P_0 = 1$ atm. On the assumption that the surface phase is mobile, the two-dimensional surface pressure π_s can be derived from the gas pressure P by means of the Gibbs equation

$$\pi_s = RT \int_0^P \theta \, \mathrm{d} \ln P$$

de Boer has assigned a value of $\pi_s = 0{\cdot}338$ dynes cm^{-1}. At this pressure a two-dimensional gas has the same intermolecular distance of 30 Å as in the gas phase. Under these conditions $\theta_s = \dfrac{A_0}{4{\cdot}08T}$ Å^2 mole^{-1}, where A_0 is the van der Waals cross-section. Rideal and Kemball proposed

22·5T Å^2 as the area per molecule in the standard state. This corresponds to a pressure of 1 atm if the thickness of the adsorption layer is arbitrarily chosen to be 6 Å. π_s then becomes 0·0608 dynes cm^{-1}. A more general method is to consider the standard state as that at which the fractional surface coverage is $\theta = 0{\cdot}5$. Utilizing the Langmuir adsorption isotherm in the form

$$\theta = \frac{Pb_T}{1+Pb_T}$$

where b_T is a function of temperature alone and equal to

$$b_T = \frac{h^3}{(2\pi m)^{3/2}(kT)^{5/2}}\frac{f_a}{f_g}e^{\Delta H/RT}$$

where f_a and f_g are the partition functions in the adsorbed and gaseous states and ΔH is the energy required to transfer the molecule from the lowest adsorbed state to the lowest gaseous state. At $\theta = 0{\cdot}5$

$$P = \frac{1}{b_T}e^{-\Delta H/RT}$$

from which b_T can be evaluated. The free energy change is given by

$$\Delta G = RT\ln\frac{P}{P_0} = -RT\log b_T P_0 - \Delta H$$

where P_0 is 1 atm, the standard pressure. The entropy change is accordingly

$$\Delta S = -\frac{\partial \Delta G}{\partial T} = R\ln b_T P_0 - RT\frac{\partial \ln b_T}{\partial T}$$

The total loss of entropy on adsorption may be regarded as separable into two parts, translational and rotational, although, as we have noted, there may be a gain in entropy due to the onset of vibrational movements; or the loss is

$$\Delta S = \Delta S_{\text{trans}} + \Delta S_{\text{rot}} - \Delta S_{\text{vib}}$$

We may consider two extreme cases. Firstly, where the adsorbed species behaves as a two-dimensional gas, there is a loss of one translational degree of freedom. The translational entropies of perfect three- and two-dimensional gases are respectively

$$\Delta S_{\text{trans},\,g} = R\ln m^{3/2}T^{5/2} - 2{\cdot}30$$

$$\Delta S_{\text{trans},\,a} = R\ln mTa + 65{\cdot}80$$

where a is the area available per molecule, an expression derived by Kemball. The rotational entropy of a molecule free to rotate in three independent ways is

$$\Delta S_{\text{rot}} = R\left[\ln\frac{1}{\pi\sigma}\left(\frac{8\pi^3(I_A I_B I_C)^{1/3}kT}{h^2}\right)^{3/2}\right]+\frac{3}{2}$$

where σ is the symmetry number and I_A, I_B, I_C are the three moments of inertia, whereas the entropy associated with each vibration frequency is given by

$$\Delta S_{\text{vib}} = R\left[\frac{h\nu}{kT}(e^{h\nu/kT}-1)^{-1}-\ln(1-e^{h\nu/kT})\right]$$

At the other extreme the molecules are regarded as immobile and suffer a complete loss of translational entropy, there may still be some rotational movement limited on strong adsorption to one degree of freedom. A vibrational movement perpendicular to the surface may still be important.

In addition to the thermal entropy there will be a configurational entropy due to the possible modes of distribution of the molecules on the surface. The total entropy of a localized phase is accordingly

$$\Delta S = \Delta S_{\text{thermal}}+\Delta S_{\text{config}}$$

where ΔS_{config} is equal to the configurations of N atoms on N_A sites

$$= \frac{N_A!}{N!(N_A-N)!} = R\ln\left(\frac{1}{\theta}-1\right)$$

and the molar configurational entropy will be

$$\Delta S_{\text{config}} = R\left[\frac{1}{\theta}\ln\frac{1}{\theta}-\left(\frac{1}{\theta}-1\right)\ln\left(\frac{1}{\theta}-1\right)\right]$$

At $\theta = 0{\cdot}5$, ΔS_{config} is zero.

Expressions have been elaborated for the entropy changes involved on adsorption when interaction between neighbours or surface association or dissociation occur. The experimental data available on physical adsorption suggest that there are many cases in which the entropy loss is less than that to be anticipated by the loss of one degree of translational freedom, suggesting that the molecules are vibrating about their vertical axes. Kiselev gives the data in Table XXIV for the adsorption of some hydrocarbons on graphite at 20°C. The loss in entropy of 1–3 e.u. is considerably less than that anticipated, i.e. $\sim 9{\cdot}0$ e.u. Hoory and Prausnitz find that on graphitized carbon the loss of entropy on adsorption is close to that anticipated for a two-dimensional gas with hindered

TABLE XXIV

Hydrocarbon	μmoles mm^{-2}	$Å^2$	ΔH at $\theta=0{\cdot}5$ (kcal $mole^{-1}$)	ΔU_0 (kcal $mole^{-1}$)	ΔS kcal deg^{-1} $mole^{-1}$
n-Pentane	3·68	45·0	10·2	−3·9	−2·3
n-Hexane	3·22	51·5	12·1	−4·5	−2·6
n-Heptane	2·89	57·3	14·0	−5·2	−3·0
Benzene	4·10	14·3	10·0	−1·85	+1·6

molecular rotation and the acquirement of vertical vibrations. Some of their data are given in Table XXV. Kemball has examined the case of benzene adsorbed on mercury taking $\sigma=6$ for the symmetry number of a molecule adsorbed with the plane of the ring held parallel to the surface and the following values for the molecular constants $M=78$, $I_A=2{\cdot}906\times10^{-38}$ g cm^2, $I_B=I_C=1{\cdot}453\times10^{-38}$ g cm^2, and $\pi=0{\cdot}0608$ dynes cm^{-1}. He found, from measurements of the surface tension over the temperature range 25–75°C, ΔG ranging from −8·51 to −7·25 kcal $mole^{-1}$ and ΔH from −16·4 to −15·7 kcal $mole^{-1}$, giving values of ΔS from −26·6 to −23·9 cal deg^{-1} $mole^{-1}$. The calculated values for the entropy of gaseous benzene are $\Delta S_{trans,g}=39{\cdot}4$ and $\Delta S_{rot}=21{\cdot}0$ e.u., giving a total of 60·4 e.u. whilst the entropy of adsorbed benzene due to restriction of one degree of translation and confinement of rotation to one in the plane of the ring is $\Delta S_{trans,a}=3{\cdot}04$, $\Delta S_{rot,a}=4{\cdot}0$, thus showing a loss of 26·0 e.u. on adsorption, in close agreement with the experimental figure.

TABLE XXV

Hydrocarbon	ΔH (kcal $mole^{-1}$)	Entropy values (cal deg^{-1} $mole^{-1}$) $-\Delta S_{trans}$	$-\Delta S_{rot}$	ΔS_{vib}	$-\Delta S$
Methane	3·02	10·67	0·06	4·04	6·69
Ethane	4·60	11·31	1·68	4·42	8·57
Propane	5·87	11·69	2·64	4·71	9·62
n-Butane	7·45	11·96	4·07	4·91	11·12

There appears to be but little difference between the physical heats of adsorption of the saturated hydrocarbon and of the olefins on this substrate.

De Boer has examined the thermal data available for the adsorption of argon, nitrogen and carbon monoxide on alumina, charcoal and graphite to evaluate the entropy changes on adsorption. For the first two gases he found that the loss of entropy on adsorption approximated to that anticipated if they behaved as two-dimensional gases but at high pressures there is a contribution to the entropy which may be accounted for on the hypothesis that there is some vibrational energy associated with the adsorbed molecules. At low temperatures, $\sim 150°K$, the loss of entropy is greater than that attributable to the loss of one degree of translation motion; this extra loss may be accounted for either by interference in the two-dimensional motion or by the freezing on to fixed sites of a small proportion of the adsorbed molecules. With carbon monoxide at room temperatures when adsorbed on alumina there is a large entropy loss for small values of θ. Thus at $\theta = 0{\cdot}003$ the gas appears to be immobile, attaining mobility at $\theta = 0{\cdot}011$ and at $\theta = 0{\cdot}1$ to $0{\cdot}55$ vibrational energy appears. We have had occasion to note that this apparent weakening of the interaction of the gas with the substrate as θ increases is found in many other cases including chemisorbed gases, e.g. hydrogen on nickel.

Orr measured the variation of the heats of adsorption of argon, nitrogen and oxygen on the crystalline alkali halides and from his values for ΔG with $\theta = 0{\cdot}5$ an entropy loss of some 16 e.u. can be calculated, in close agreement with that anticipated for a two-dimensional gas. The large changes in entropy associated with strong adsorption on fixed sites are noted in the chemisorption of nitrogen on iron and of water on zinc oxide as well as in the physical adsorption at low temperatures of hydrogen on glass and on chabazite. In the case of the adsorption of hydrogen on metals it would appear that at low temperatures the adatoms are not mobile but mobility increases both with temperature and surface coverage. Even in physical adsorption of the rare gases on metals at low temperatures heterogeneity of sites may be detected. Thus, over the range 77–90°K the heats of adsorption of xenon and krypton on nickel were found to fall from 5·4 kcal mole^{-1} to 4·5 kcal mole^{-1} for krypton and 4·75 kcal mole^{-1} to 4·60 kcal mole^{-1} for xenon by making use of the Clausius-Clapeyron equation

$$\Delta H_i = RT^2\left(\frac{\partial \ln P}{\partial T}\right)_\theta$$

On evaluating the molecular entropies over the range of pressure it was found that the data were best interpreted on the hypothesis that there were two uniform adsorbing areas on the nickel each with a heat of adsorption conforming to

$$\Delta H_{i\theta} = \Delta H_{i\theta_0} + A\theta$$

which for krypton gave: Patch I, $\Delta H_i = 4{\cdot}05$ kcal mole^{-1}, $\theta = 0{\cdot}24$; Patch II, $\Delta H_i = 5{\cdot}10$ kcal mole^{-1}, $\theta = 0{\cdot}76$.

Formal Reaction Kinetics

A study of the influence of pressure changes in reactants or products on the reaction kinetics frequently gives us some clue as to the mechanism of the reaction. This can be treated in a formal manner on the assumption that the rate of reaction is proportional to the amount on the surface and that this amount is governed by a Langmuir isotherm. For simple thermal decompositions in which the adsorbate occupies but one elementary site, we have

$$\omega = k\theta = k\frac{bP}{1+bP}$$

where k is the velocity constant of the reaction. If the temperature of the reaction is varied we have

$$\frac{\partial \ln k}{\partial T} = \frac{E}{RT^2}, \quad \frac{\partial \ln b}{\partial T} = -\frac{\lambda}{RT^2} \text{ and } \frac{\partial \ln \omega}{\partial T} = \frac{E-\lambda}{RT^2}$$

where E is energy of activation of the process, and λ the heat of adsorption of the reactants or latent heat of evaporation. Dependent on the strength of binding of the reactant to the substrate the kinetics can evidently range from zero to first order on progressive weakening of the bond. Intermediate values in the Langmuir range have frequently been equated to the rate expression

$$\omega = kbP^{1/n}$$

In many cases the products of decomposition act as inhibitors. Here again it can be assumed that the products are also adsorbed in accordance with the Langmuir equation. If the reactant is weakly adsorbed and the product is a strong inhibitor, the reaction kinetics follow the expression

$$\omega = k\frac{bP}{b'P'}$$

where b', P' represent the Langmuir constant and the partial pressure of the product. The apparent and true energies of activation are accordingly given by the relationship

$$\frac{\partial \ln \omega}{\partial T} = \frac{\partial \ln k}{\partial T} + \frac{\partial \ln b}{\partial T} - \frac{\partial \ln b'}{\partial T}$$

or

$$E_{\text{apparent}} = E_{\text{true}} - \lambda_{\text{react}} + \lambda_{\text{prod}}$$

The decomposition of alcohols on alumina results in the formation of acetaldehyde and water, the latter being a strong inhibitor. The observed energies of activation for this decomposition differ markedly at high and at low pressures. Table XXVI shows some values that have been recorded. The difference between the two values, about 13·8 kcal mole^{-1},

TABLE XXVI

Alcohol	E (kcal mole^{-1}) High pressure	Low pressure	Difference
n-Propyl	41	28·5	12·5
Isopropyl	39	26	13
n-Butyl	44	28	16

represents the heat of adsorption of water on alumina. More complex are those cases where it is assumed that both reactant and product are strongly adsorbed. The formal equation for the reaction rate would be

$$\omega = k\frac{bP}{1+bP+b'P'}$$

and the decomposition of ammonia on a tungsten surface has been claimed to follow such kinetics, but other experimental data support the view that two adsorbable gaseous products are formed such as nitrogen and hydrogen. If such is the case we have

$$\omega = k\frac{bP}{b'P'+b''P''}$$

where P' and P'' are the partial pressures of the two gaseous products. It is possible to treat bimolecular surface reactions in a similar manner. Here we must distinguish between two possibilities. In the first, the reaction can be assumed to occur between a chemisorbed species and one from the gas phase, either from a van der Waals layer or by direct impact. The other case is one in which reaction takes place between two chemisorbed species. In the former case $\omega \propto P_2\theta_1$ and in the latter $\omega \propto \theta_1\theta_2$, where subscripts 1, 2 refer to the two reactants. The view that reaction takes place only when the two reactants are adsorbed on neighbouring sites requires further consideration if what is in effect a two-dimensional gas reaction is really involved. In general, both of these types of reaction should conform to the Langmuir formulation

$$\omega = k\frac{b'P'}{1+b'P'+b''P''}\times\frac{b''P''}{1+b'P'+b''P''}$$

on the assumption that the Langmuir adsorption isotherm is applicable to both reactants whether physically adsorbed or chemisorbed. The expression can be greatly simplified as Hinshelwood pointed out when assumptions are made as to the ease of adsorption of reactants and products. Thus, if both reactants are weakly adsorbed and the products not at all, the reaction rate can be written

$$\omega = k \times b'P' \times b''P''$$

This relationship obtains over certain limits of temperature and pressure for the combustion of hydrogen and oxygen on porcelain, for the hydrogenation of ethylene on copper, for the hydrogenation of benzene on nickel, and the bromination of ethylene on glass. If one reactant is strongly adsorbed and the second one somewhat weakly, the two mechanisms are indistinguishable and the equation reduces to

$$\omega = k\frac{b'P'b''P''}{(1+b'P')^2}$$

which attains a maximum when $P' = 1/b'$. Such a relationship is found for the water gas shift reaction $CO_2 + H_2 \rightleftharpoons CO + H_2O$ on platinum. Further, if one reactant is weakly adsorbed and the second one much more strongly, we obtain

$$\omega = k\frac{b'P'}{b''P''}$$

Within certain limits, this relationship obtains for the oxidation of carbon monoxide and of hydrogen on platinum, and of carbon monoxide on quartz, for the hydrogenation of ethylene on copper over a limited range and for the interaction of hydrogen and nitrous oxide on platinum. Although these formal relationships are readily obtained on the simple premises which have been outlined, they do not assist us to a marked extent in unravelling the actual molecular mechanism. Further insight is obtained by examining Langmuir's concepts more closely. The chemi-adsorption of a gas according to his view is similar to that involving bond formation in ordinary chemical reactions. Thus we may have simple site occupation as an M—H or two site occupation as in $\begin{matrix} M \\ M \end{matrix}\!\!>\!SO_2$.
Again, in the adsorption of diatomic gases such as N_2 or H_2 we may have surface compounds formed such as M—H_2, $\begin{matrix} \text{M—N} \\ \;\;\;\,\| \\ \text{M—N} \end{matrix}$ or M—H. It can readily be shown, as in the case of dissociation, that different

surface valencies of reactants or products can also give rise to different functional relationships between pressure and reaction velocity. We thus note by making somewhat arbitrary assumptions as to the surface valency and the molecular kinetics the dependence of the reaction rates of the catalytic reactions on the pressure can be satisfactorily interpreted, indeed the cases of surface dissociation which necessitate a square root dependence receive a more satisfactory explanation than one based on the assumption of a uni- or bimolecular reaction between molecules of the reactant whose surface concentration is governed by a simple Langmuir adsorption isotherm. In all these reactions the rate is governed by the reaction rate involving the chemisorbed species. We have noted, however, that in chemisorption energies of activation may be involved and for some reactions it is the energy of activation for chemisorption of one of the reactants which governs the reaction velocity. H. S. Taylor (Spiers Memorial Lecture, *Transactions of the Faraday Society*, 1950) cites the rate of decomposition of ammonia on a doubly promoted iron catalyst as a typical example. The kinetics of this reaction can be formulated

$$\frac{\partial x}{\partial t} = k\frac{P_{NH_3}^{0\cdot6}}{P_{H_2}^{0\cdot9}}$$

suggesting a strong adsorption of one of the products, hydrogen, and a moderate or dissociative adsorption of the ammonia. He points out that the actual rate-determining reaction is the desorption of the nitrogen, its activated adsorption being the controlling process in ammonia synthesis.

In addition to following the rates of reaction as a function of the partial pressures of reactants and products, the use of radio isotopes for tagging a reactant or product can provide further information.

On Absolute Reaction Rates

In order to evaluate the absolute rate of a reaction, i.e. the rate of reaction per square centimetre of catalytically active surface, we may treat the Arrhenius equation $k = A\,e^{-E/RT}$ by kinetic or transition state methods, of which the former is of more limited applicability.

Kinetic treatment

It is customary to consider the non-exponential term in the Arrhenius equation as containing a collision number Z and a "steric factor" P which latter represents the probability that reaction occurs when all the conditions for reaction are fulfilled and that the collision has the necessary energy of activation E. The equation then becomes

$$k = \mathrm{P}Z\,\mathrm{e}^{-E/RT}$$

Although in some gas reactions the P term is found to be unity, there are many exceptions and in surface reactions we have but little information as to what must be considered as important in evaluating this term. A case for which a kinetic treatment is applicable is one in which we consider the condensation or reaction of a substance on impact with the surface. The rate of reaction is then determined by the Herz-Knudsen expression

$$\omega = \mathrm{P}n\left(\frac{kT}{2\pi m}\right)^{1/2}\mathrm{e}^{-E/kT}$$

where n is the concentration of the reactant in molecules cm^{-3} and m its molecular mass. For oxygen at 300°K and at a pressure of 1 mm Hg, ω becomes

$$\omega = \mathrm{P}\times 6{\cdot}0\times 10^{20}\,\mathrm{e}^{-E/RT}\ \text{molecules cm}^{-2}\,\text{sec}^{-1}$$

Since there are only some 10^{14}–10^{15} sites per square centimetre we note that unless the energy of activation for reaction is very large the rate of condensation or reaction becomes exceedingly fast. For nitrogen chemisorbed on iron the value of $\mathrm{e}^{-E/RT}$ is some e^{-8} which would result in a rate of some 10^{12} molecules $\mathrm{cm}^{-2}\,\mathrm{sec}^{-1}$, of the order found by experiment.

There are a number of catalytic surface reactions in which the mechanism suggested is that of a two-dimensional gas reaction operative between the mobile chemisorbed species. The simplest case is that concerning the recombination of atoms adsorbed on the surface to form molecules which subsequently evaporate. If the adatoms are regarded as behaving as a two-dimensional gas, the rate of atomic collision and thus the rate of molecular evaporation can be calculated. The total number of collisions made per cm^3 per second between the atoms of concentration n and mass m and target diameter σ is given by

$$Z = 4\pi n^2\sigma^2\left(\frac{kT}{2\pi m}\right)^{1/2}$$

The volume associated with 1 cm^2 of catalyst surface is regarded as having a depth of the order of σ the atomic diameter. The number of collisions in this volume having regard to the fact that only those collisions in the plane of the surface are considered, i.e. two-thirds of the possible collisions in volume, is consequently

$$\tfrac{4}{3}n^2\,\sigma\left(\frac{2\pi kT}{m}\right)^{1/2}$$

and the rate constant of the reaction becomes

$$k = \mathrm{P}\frac{4\sqrt{2}}{3}n^2\,\sigma\left(\frac{\pi kT}{m}\right)^{1/2}\mathrm{e}^{-(2E-D)/RT}$$

where E is the desorption energy for the atoms and D is the molecular dissociation energy.

If we consider the case of oxygen on tungsten where the surface population is $\sim 10^{14}$ atoms at 300°K and the target diameter $\sigma = 4\times 10^{-8}$ cm with an atomic mass $m = 2{\cdot}67\times 10^{-1}$ g, we obtain

$$k = P\times 5{\cdot}3\times 10^{25}\,\mathrm{e}^{-(2E-D)/RT}$$

Thus, an apparent activation energy of some 4 kcal mole^{-1} is required to bring the observed reaction velocity into agreement with that calculated.

Transition state treatment

The concept of the critical complex introduced by Marcelin has been developed by Polanyi, M. G. Evans and Eyring and applied to reaction kinetics and to heterogeneous catalytic actions in particular by Laidler. Attention is focused on the act of collision of the reacting molecules. It is assumed that on collision a transitory complex is formed. The complex is regarded as containing the energy lost in collision and is thus an "activated complex" situated at the top of an energy barrier. The rate of reaction is essentially the rate at which the activated complex travels over the barrier. It may revert to the original constituents, and in consequence the concept of a transmission coefficient κ has been introduced to indicate the efficiency of the formation of the products. In general, the transmission coefficient κ is taken as unity. The assumption is made that during reaction the concentration of the activated complex is in equilibrium with the reactants. In the application of this concept to heterogeneous catalytic action the surface sites are regarded as a reactant; consequently, there exists an equilibrium between the reactant concentration, the surface sites and the activated complex of surface site and reactant. If the concentration of surface sites is C_s, of reactant is C_r, and if C_a is the surface-activated complex, we have an equilibrium

$$\frac{C_a}{C_s C_r} = K$$

We may accordingly write

$$K = \frac{f_a}{f_s f_r} e^{-E/RT}$$

where f_a, f_s and f_r are the partition functions of the reactants and product, and E is the zero point energy or the energy increase at 0°K when one gram molecule of such reactant forms a gram molecule of product, all being in their standard states. The difference in zero point energies of reactants and transition complex in the standard state approximates to the energy of activation at room temperature.

The partition functions can be factorized into translational, rotational and vibrational factors. In considering the decomposition of the activated complex, one vibrational bond undergoes rupture to yield the products of the reaction. Thus, in the partition function of the activated complex, f_a, one vibrational bond exists in which ν tends to zero: the factor $(1 - e^{-h\nu/kT})^{-1}$ is thus replaced by $\operatorname{Lt}_{\nu=0} (1 - e^{-h\nu/kT})^{-1} = kT/h\nu$ and the equilibrium constant K becomes

$$K = \frac{kT}{h\nu} \frac{f'_a}{f_s f_r} e^{-E/RT}$$

where f'_a, having one degree less of vibrational freedom, replaces f_a. We thus obtain

$$\nu C_a = C_s C_r \frac{kT}{h} \frac{f'_a}{f_s f_r} e^{-E/RT}$$

The rate of reaction is $\omega = \nu C_a$ and with kT/h, which is about $1{\cdot}46 \times 10^{13}$ at 700°K, is thus obtainable from the partition function and concentration of reactants. We may consider the cases of decomposition on impact on a catalytic surface both when the activated complex is immobile and when it is free to move. If the layer is immobile there is no translational energy contribution by the activated complex, the contribution of the vibrational and rotational energies we can term b_a. The sites are considered to have only vibrational energy and we may write $f_s = 1$, i.e.

$$\omega = \frac{kT}{h} \frac{C_s C_r b_a}{f_r} e^{-E/RT}$$

The partition function for the gas may be split up into three translational degrees of freedom $(2\pi mkT/h^3)^{3/2}$ and the vibrational and rotational b_g. Accordingly, we have

$$\omega = \frac{kT}{h} C_s C_r \frac{b_a}{b_g \left(\frac{2\pi mkT}{h^3}\right)^{3/2}} e^{-E/RT}$$

Since the activated complex is immobile and has no translational moment and is assumed to have no rotation, b_a is only vibrational and has a small partition function equal to 1 apart from its symmetry factor σ_a. Similarly, the vibrational part of $b_g = 1$ and only the rotational part $8\pi^2 IkT/h^2 \sigma_g$ counts. Hence

$$\omega = \frac{\sigma_g}{\sigma_a} C_r C_s \frac{h^4}{8\pi^2 I (2\pi mkT)^{3/2}} e^{-E/RT}$$

The data in Table XXVII have been calculated by Laidler ("Chemical Kinetics", p. 168) for the catalytic decomposition of various gases.

TABLE XXVII

Gas	Catalyst	Temp. (°K)	Rate (sec^{-1}) calc.	obs.
PH_3	Glass	684	$2{\cdot}2 \times 10^{-8}$	$4{\cdot}7 \times 10^{-7}$
HI	Platinum	836	$1{\cdot}2 \times 10^{-3}$	$1{\cdot}0 \times 10^{-3}$
N_2O	Gold	1211	$3{\cdot}4 \times 10^{-6}$	$12{\cdot}3 \times 10^{-5}$

If the adsorption layer is mobile, we consider the activated complex to be the molecule just as it strikes the surface. Hence

$$K_c = \frac{C_a}{C_r} = \frac{f_a}{f_g} e^{-E/RT}$$

and

$$\omega = \frac{kT}{h} C_r \frac{f_a}{f_g} e^{-E/RT}$$

Since the activated complex is still gaseous, it will have vibrational and rotational degrees of freedom like the other molecules and f_a/f_g reduces to the ratio of the translational terms of which there are three in the gas phase and two in the surface, i.e. $(2\pi mkT/h^3)^{3/2}$ and $2\pi mkT/h^2$, respectively. Hence

$$\omega = \frac{kT}{h} C_r \frac{h}{(2\pi mkT)^{1/2}} e^{-E/RT} = C_r \left(\frac{kT}{2\pi m}\right)^{1/2} e^{-E/RT}$$

We note that $C_r = \mathrm{P}n$ and when the energy of activation becomes zero the expression reduces to the Hertz-Knudsen relationship

$$\omega = \frac{\mathrm{P}}{(2\pi mkT)^{1/2}}$$

Laidler and his colleagues have developed transition-state velocities of reaction for many other modes of hypothetical surface reactions.

Treating the transition state complex as in mass equilibrium with the reactants we have for the equilibrium constant between complex and reactants

$$K = \frac{C_a}{C_s C_r}$$

We can assign a standard free energy of activation ΔG and an entropy of activation ΔS to the formation of the complex thus

$$-\Delta G = T\Delta S - \Delta H = RT \ln K$$

The reaction rate constant is evidently $k = \bar{k}K$ where $\bar{k}$ is the specific reaction rate; thus

$$k = \bar{k}\,\mathrm{e}^{\Delta S/R}\,\mathrm{e}^{-\Delta H/RT}$$

which may be compared with the Arrhenius equation

$$k = A\,\mathrm{e}^{-\Delta E/RT} = \mathrm{P}Z\,\mathrm{e}^{-\Delta E/RT}$$

If we identify the energy of activation of the transition complex with the Arrhenius energy of activation we have

$$\bar{k}\,\mathrm{e}^{\Delta S/R} = \mathrm{P}Z$$

and making the assumption that only one vibrational bond is involved in the activated complex

$$\bar{k} = \frac{kT}{h}$$

or

$$\mathrm{P}Z = \frac{kT}{h}\,\mathrm{e}^{\Delta S/R}$$

Thus we find that a substantial increase in entropy in forming the transition state complex is reflected in the P term or a high probability of reaction.

Chapter 5

Catalytic Oxidation

Catalytic oxidation reactions are of importance in many industrial processes. These include the oxidation of carbon monoxide, sulphur dioxide, ammonia, methanol and hydrocarbons such as naphthalene as well as processes involving the peroxidation and epoxidation of olefins.

As heterogeneous catalysts both metals and metallic oxides have been used, and information pertinent to any views as to the mechanism of the catalytic action can be obtained from studies on the interaction of oxygen with them.

Metals

The low-temperature oxidation of metals and effects of corrosion have been investigated in some detail and the theoretical treatment has been developed by Mott for oxide formation and by U. R. Evans in corrosion. These effects, due to oxygen penetration in depth of oxides in granular form and of different molecular volumes, have little bearing on surface catalysis. It may, however, be noted that electron transport is rate controlling in some cases, e.g. the low-temperature oxidation of tungsten and of zinc, and mass transport in other cases, e.g. the high temperature oxidation of these metals. With copper, nickel, bismuth and tin the metal is more mobile than the oxygen. With tungsten, zinc and titanium the reverse is the case.

If a surface compound M_nO is formed, on elevation of the temperature it decomposes

$$M_nO \rightarrow M_n + O$$

$$2M_nO \rightarrow 2M_n + O_2$$

or evaporates as a compound, e.g. MO or MO_2. The choice will evidently depend on the strength of the M—O bond relative to that binding the M_nO complex to the substrate. Only the higher oxides and halides formed at metallic surfaces are volatile for the bond between substrate and compound becomes progressively weaker as the ease of charge transfer increases.

Hot Wire Reactions

The hot wire technique permits us to study the rate of oxygen removal and the nature of the products formed.

If a platinum wire is heated in oxygen there is a fall in pressure; the kinetics of the reaction and analysis of the products show that the oxygen is condensed on the walls as an oxide PtO_2. This is formed by two reactions, one involving formation from evaporating platinum atoms and gaseous molecular oxygen and one by direct reaction of oxygen with the platinum surface to form a volatile oxide PtO_2. The rate of this reaction is proportional to P_{O_2} and E is ~ 63 kcal mole^{-1}. We can compare this with other "wire" reactions: thus a similar diminution of pressure is noted in nitrogen, but here the fall is not due to compound formation but to occlusion of nitrogen in the deposited platinum which is acting as a "getter".

On tungsten a surface oxide WO is formed with $E = 23$ kcal mole^{-1}; this can subsequently undergo thermal decomposition to yield oxygen atoms ($E = 160$ kcal mole^{-1}). The species observed in the field emission microscope as migratory is probably WO and not O. The metal is thus an active catalyst for the dissociation of oxygen, and behaves in a manner analogous to its behaviour in hydrogen.

In iodine, likewise, a surface compound, WI is formed which subsequently decomposes to liberate iodine atoms. Tungsten is not attacked by chlorine, although atomic chlorine readily attacks the filament. Iron on evaporation will remove nitrogen as FeN_2.

When platinum is exposed to iodine below 1400°K, the metal is not attacked but acts as a catalyst for the dissociation to atomic iodine. At higher temperatures, atomic iodine attacks platinum ($E = 53$ kcal mole^{-1}). PtI evaporates and in its passage to the walls may react with iodine atoms to condense as a mixture of PtI and PtI_2. In these high-temperature reactions the energy of activation may be regarded as the energy required for electron transfer uncomplicated by any subsequent process of evaporation. We can thus write

$$E = e[\phi_0 - x_1 + x_2 - \phi_a + \Delta V]$$

where ϕ_0 is the work function of the metal, ϕ_a is the electron affinity of the oxygen and x_1, x_2 are the work functions of the oxide–metal and oxide–gas interfaces respectively. ΔV is the dipole contribution of the adsorbed oxygen to the work function of the metal. If during reaction the fraction of the surface covered with oxide is small x_1 and x_2 can be neglected, and in any case the difference between x_1 and x_2 is small. Table XXVIII gives the data for attack by oxygen. Any metal with a

TABLE XXVIII

Metal	ϕ_0 (eV)	ϕ_a (eV)	ΔV (eV)	E (eV)	E_{exp} (eV)
Pt	6·35	3·4	0·2	2·75	2·75
W	4·65	3·4	0·2	0·94	0·88
C	4·31	3·4	0·2	0·7	0·52

low work function, i.e. $\phi_0 < 3{\cdot}49$ eV, should on this basis react with oxygen without any energy of activation. In a similar manner the data on the catalytic decomposition of ammonia on metal surfaces (which in all probability involves one M—N bond) are found to support a simple electron transfer mechanism (Table XXIX). When the surface oxide is not volatile it is possible to obtain some information by electron diffraction. Nickel has been examined in this manner. It would appear that molecular surface adsorption precedes surface dissociation; the molecular species has surface mobility and can travel to lattice defects and boundaries where dissociation and chemisorption take place.

TABLE XXIX

Metal	ϕ (eV)	E_{exp} (eV)	$\phi - E$ (eV)
Pt	6·35	3·5	2·85
W	4·65	1·7	2·95
Mo	4·36	1·39	2·97
Ni	4·05	1·13	2·92

This process can extend to give a surface lattice with a structure corresponding in one plane to a surface composition Ni_3O. This in its turn is converted into a second surface lattice type in which certain of the nickel atoms are replaced by oxygen. Thus, before oxidation proceeds in depth the composition of the surface layer is not uniform but, depending on the extent of the [100], [110] and other planes, patches of metal as well as surface oxides of composition from Ni_3O to NiO can co-exist. Magnetic measurements suggest that, when excess oxygen is present, Ni^{3+} ions are formed, rendering the catalyst ferro-magnetic. Thus, from the point of view of valency, we have different levels of oxy-redox potential governed by

$$\frac{Ni^{0}}{Ni^{+1}}, \frac{Ni^{+1}}{Ni^{+2}}, \frac{Ni^{+2}}{Ni^{+3}}$$

with corresponding changes in the height of the Fermi level.

Metal Oxides: Oxygen Exchange

Winter and Boreskov have examined the process of interaction of oxygen with oxides with the aid of isotopic oxygen $^{16}O^{16}O$, $^{16}O^{18}O$, and $^{18}O^{18}O$. The adsorption rate in many cases follows that anticipated in the Elovitch equation for an activated dissociative process

$$\frac{d\theta}{dt} = KP_{O_2}(1-\theta)^2 e^{-(E_a-\alpha\theta)/RT}$$

Since oxygen has a high electron affinity, the adsorbed form can be regarded as predominantly O^- or O^{2-} and adsorption can take place with ease and most readily on p-type semiconductors. It has been established that chemisorbed oxygen can equilibrate on the surface, i.e. adsorption of $^{16}O_2$ and $^{18}O_2$ results in the desorption of $^{16}O^{18}O$ in the gas mixture. At somewhat higher temperatures exchange can also take place with the lattice oxygen of the oxide

$$^{16}O_2 + M^{18}O \rightarrow M^{16}O + ^{16}O^{18}O$$

Although in general gas–surface and surface exchanges take place more readily than exchange with lattice oxygen, the latter rate is greatly influenced by the treatment which the catalyst has undergone, i.e. is greatly dependent on the magnitude of its p-type characteristics. Zinc oxide as well as nickel oxide can be rendered active to exchange at room temperature by suitable doping.

Isotopic oxygen exchange can take place on metal as well as on oxide surfaces. On platinum and also on Co_3O_4 exchange can take place at temperatures as low as 200°C, rising to some 500°C on vanadium and chromium oxides.

There appears little doubt that oxygen in the form of oxide participates in the exchange reaction. The bonding energy of oxygen chemisorbed by the oxides is very close to, if not identical with, that of the oxygen in the surface layer of the crystal of the oxide.

Two possible mechanisms can be formulated

$$^{18}O_a + ^{16}O_m \rightarrow ^{16}O^{18}O(g) \tag{a}$$

$$^{16}O^{16}O(g) + ^{18}O_a \rightarrow ^{16}O^{16}O^{18}O_a \rightarrow ^{16}O^{18}O(g) + ^{16}O_a \tag{b}$$

where O_a represents oxide oxygen or oxygen chemisorbed on a lattice defect and O_m represents migratory oxygen atoms. Reaction (b), a Rideal mechanism, is that favoured by Boreskov.

Since the lattice exchange and surface exchange rates are comparable, it is to be anticipated that the metal–oxygen bonding energy is important in catalytic oxidative processes. Amongst the oxides those of the transition metals with 3, 4, 8 and 9 d-electrons in the cations are most active and the heats of conversion of higher to lower oxides, e.g. $V^{5+} \rightarrow V^{4+}$ or $Co_3O_4 \rightarrow CoO$, give a good guide to the specific catalytic activities.

It is possible to make an estimate from spectral data of the energy consumed when a $2p$-level in oxygen receives an electron from a lower $3d$-level in various metallic oxides. These energies vary from 0·67 eV for CoO to over 3·0 eV in TiO_2 and ZnO. It is interesting to note that in the periodic table two maxima are obtained for the metallic oxides and that maxima and minima are to be found for a great variety of catalytic oxidations such as of CO, NH_3 and hydrocarbons and for the decomposition of H_2O_2, N_2O and ozone. In addition, they are also found in the catalytic combination of oxygen atoms and isotopic oxygen exchange reactions. The minimum between can be attributed to the stability of the $3d^5$ configuration.

The Hydrogen–Oxygen Catalysis

The catalytic formation of water at metallic and oxide surfaces has not been the subject of detailed investigation since reproducibility is difficult to achieve.

The rates of oxidation of hydrogen in excess oxygen on various oxides are according to Boreskov directly parallel with the rate of isotopic exchange of surface and lattice oxygen with the molecular form.

Charcoal as Oxidation Catalyst

Although charcoal is generally employed as an adsorbent, it has been used as a model to simulate biological oxidases (chiefly by Warburg, Meyerhof and Freundlich). Active charcoals possessing large specific surfaces, both macro- and microporous, can be prepared by treatment with steam during combustion or ignition with salts such as zinc chloride. A surprisingly active charcoal can be derived from sugar containing 10 % urea and burning the resulting charcoal until some two-thirds of it have been removed by combustion. A specific surface of 200 $m^2\ g^{-1}$ is readily obtained. The surface of the carbon is composite, partly graphite capable of π-bonding, partly aliphatic chain and partly ethenoid, with varying degrees of unsaturation. The presence of free valencies can be established by n.m.r. spectroscopy, and the heterogeneity of the surface can be established by examining the contours of the curves for adsorption or for heat liberation using various gases, especially oxygen.

Suspensions of carbon in water can serve as both hydrogen and oxygen electrodes. The former is readily reversible, especially if, as Frumkin showed, a small quantity of platinum is added to the carbon. At the same time it can undergo autoxidation at ordinary temperatures. The reaction can be written

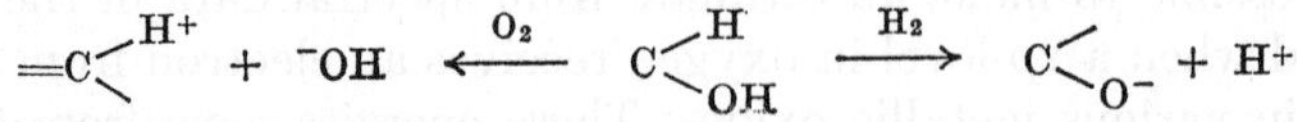

The autoxidation is found to be a zero order reaction over a wide range of oxygen pressures, with a temperature coefficient of $k_{50}/k_{40} = 1{\cdot}8$. The autoxidation process is readily poisoned by adsorption of capillary active substances such as amyl alcohol and many organic and inorganic salts. Of these, potassium ferricyanide is particularly active and selective; from adsorption data it is found that some 0·4 % of the available surface is autoxidizable. The autoxidation does not appear to be autocatalytic; this would seem to suggest that carbon chain combustion alone is involved.

Oxygen can be present on a charcoal surface in three different forms; one is removable only as CO_2 at high temperatures (1000°C); another is removable as oxides of carbon by evacuation at ordinary temperatures; the third is removable as oxygen at low temperatures. It is this last form that is catalytically reactive and can show peroxide characteristics when acidified with sulphuric acid.

A great variety of organic compounds in dilute aqueous solution can be oxidized in the presence of charcoal. In general, compounds containing two adjacent polar groups, such as the amino, oxalic or malonic acids, are more readily oxidized than alcohol or acetic acid. In the case of the oxidation of oxalic acid, the carbon dioxide evolution during oxidation is equivalent to the acid consumed, according to the equation

$$2(COOH)_2 + O_2 \rightarrow 4CO_2 + 2H_2O$$

The process of autoxidation is therefore completely inhibited. Malonic acid on oxidation consumes oxygen in the ratio $CO_2{:}O_2{::}1{:}1$, forming glyoxylic acid

$$2CH_2(COOH)_2 \xrightarrow{O_2} 2CHOH(COOH)_2 \xrightarrow{O_2} 2CHO.COOH + 2CO_2 + 2H_2O$$

The oxidation velocity rises to a well-defined maximum as the oxalic acid concentration is increased, occurring at some 0·0075 M in air at 40°C and 0·015 M in oxygen under the same conditions.

Following Langmuir, we may express the reaction rate as a function of the surface covered with oxygen atoms and oxalic acid respectively, thus

$$\omega = \frac{k\sqrt{p_{O_2}}C_{ox}}{(1+\alpha\sqrt{p_{O_2}}+\beta C_{ox})^2}$$

This is the result of competitive adsorption of the reactants. The two end portions of the ω, p_{O_2}, C_{ox} curves, viz. $\omega = k'' C_{ox}\sqrt{p_{O_2}}$ for small concentrations of oxalic acid and $\omega = k''\sqrt{p_{O_2}}/C_{ox}$ at high concentrations, are readily confirmed experimentally.

The fraction of the surface which has catalytic properties is far greater than the portion which is autoxidizable as determined by the extent of adsorption of various inhibitors. Of the various inhibitors, both caproic acid and amyl alcohol are found to be effective but they are not completely selective in that the concentrations required for inhibition vary with the nature of the capillary active reactant. It would appear that some 40% of the active charcoal surface possesses catalytic activity.

Very active catalysts may be prepared from charcoal obtained from the ignition of sugar–ferric chloride mixtures, and mixtures of urea, sugars and ferric chloride lead to even better catalysts. The Fe–C, and Fe–C–N complexes are selectively poisoned by cyanides and their specific activities for the oxidation of oxalic or other acids are readily determined. The three active areas, viz. the C–C, Fe–C and Fe–C–N portions, are differentiated by breaks in the curve relating catalytic activity to adsorbed inhibitor. Here again no inhibitor was found to be completely selective: thus, potassium thiocyanate was more selective on the Fe–C–N complex and potassium cyanide more selective on the Fe–C complex. The data in Table XXX were obtained by Miss Wright for the oxidation of oxalic acid on promoted charcoals.

TABLE XXX

Poison	Surface	Area (10^4 cm^2 g^{-1})	% total surface	% total activity	O_2 (mm^3 mg^{-1} h^{-1})	Specific activity (cm^3 O_2 cm^{-2} h^{-1})
KCNS	Fe–C	1·8	6	13	1·57	$8{\cdot}7 \times 10^{-5}$
KCNS	Fe–C–N	1·35	4·5	86·5	10·4	$7{\cdot}7 \times 10^{-4}$
KCNS	C–C	6·0	20·0	0·35	0·025	$0{\cdot}7 \times 10^{-6}$
KCN	Fe–C	1·8	6·0	6·0	0·72	4×10^{-5}
KCN	Fe–C–N	3·0	10·0	40·0	4·80	$1{\cdot}6 \times 10^{-4}$

Similar variations in catalytic activity of the promoted centres were found for the oxidation of formic and aminoacetic acids. The relative rates of catalytic oxidation of oxalic, formic, malonic and aminoacetic acids were found to fall in the order of their depolarizing action on an oxygen–platinum electrode.

Coupled Oxidation

The oxidation of copper in the absence and presence of hydrogen has been the subject matter of several investigations. The metal oxidizes readily to form a superficial oxide. It is found that the oxide formed corresponds in composition to cuprous oxide up to 175°C above which cupric oxide appears as the main constituent and the reaction proceeds most rapidly at the metal–oxide interface. The rate of oxidation is increased by the presence of hydrogen and the maximum rate is attained at a ratio $H_2:O_2::1:1$. When the rates of oxidation are compared with the rates of reduction of the resulting oxide at the same partial pressures, it is found that the rates are comparable above 175°C, i.e. the catalysis is a typical intermediate-compound reaction involving the alternate oxidation and reduction of the metal. Below 175°C more oxygen is consumed than that used for the production of water, and the experimental data suggest that in the limit equal quantities of oxygen are consumed by the hydrogen and by the metal.

Many examples of coupled oxidations are known where, in the presence of water, the oxidation of some metals results in hydrogen peroxide formation (or the oxidation of a suitable acceptor). These results may be formulated by the reaction scheme.

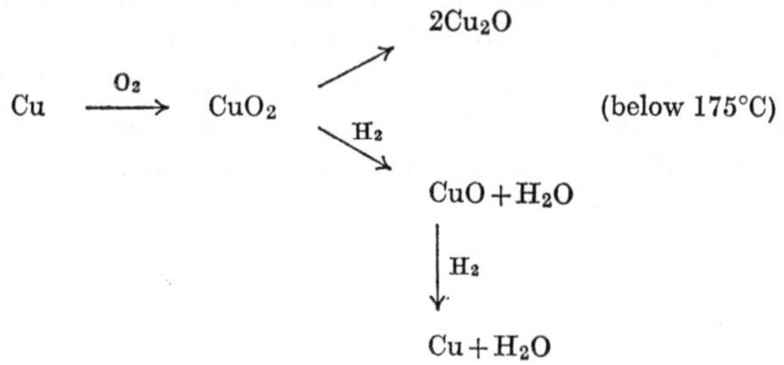

The energy of activation for the process of coupled oxidation is relatively small, being of the order of $E = 6{\cdot}0$ kcal mole^{-1}, whereas that for the catalytic process, i.e. that involving the reaction $CuO + H_2 \rightarrow Cu + H_2O$ is some $E = 9{\cdot}8$ kcal mole^{-1}. There is some uncertainty with respect to this value of $E = 9{\cdot}8$ kcal mole^{-1} since if the cupric oxide layer is very thick the reduction process involves migration of the hydrogen through a layer of copper and the activated adsorption of hydrogen on copper involves $E = 9{\cdot}8$ kcal mole^{-1} and its activated diffusion through copper $E = 14{\cdot}1$ kcal mole^{-1}.

Examination of the catalytic activity of gold wires as well as gold–copper alloys consisting of both ordered and disordered Cu_3Au, Cu_4Au

and Cu_9Au revealed the interesting fact that gold was catalytically inert for the hydrogen–oxygen combination up to 360°C and that on all the alloys there was no catalytic activity until the copper had been oxidized. If the high copper containing alloys are first oxidized, copper oxide separates and can on reduction exert a catalytic activity.

Further information on the mechanism can be obtained by examination of the combustion of hydrogen to which some carbon monoxide has been added. Hence a pronounced selectivity can be obtained. Again it is noted that below 175°C there is a tendency for the carbon monoxide to be selectively oxidized and above 175°C the hydrogen is preferred. The selectivity which may be measured by the ratio CO/H_2 consumed is dependent both on the catalyst and on the temperature, as is to be noted in Table XXXI, where the % combustible in oxygen is given.

TABLE XXXI

Catalyst	Gas composition (%) CO	H_2	Temperature (°C)	CO/H_2 oxidized
Fe_2O_3	9	77	220	4·3:1
CuO	14	74·5	280	0·466:1
CuO + 0·05% Pd	10	80	133	1·10:1
CuO + 0·05% Pd	10	80	150	1·71:1
CuO + 0·05% Pd	10	80	179	2·32:1
CuO + 2·0% Pd	10	80	150	1:1·68
CuO + 2·0% Pd			204	1:2·04
CuO + 2·0% Pd			250	1:4·16

The order of catalytic activity for the preferential combustion of the carbon monoxide is found to be Fe_2O_3 > promoted Fe_2O_3 > CuO > NiO > Pd and the kinetics of the two reactions take the form

$$\frac{\omega_{CO}}{\omega_{H_2}} = \frac{p_{CO}}{p_{H_2} p_{O_2}^{1/2}}$$

Thus the mechanism of the oxidation of the carbon monoxide must differ from that of the hydrogen since the two net reactions must be either

$$\text{(a)}\quad H_2 + \tfrac{1}{2}O_2 \rightarrow H_2O$$
$$CO \rightarrow CO_2 \quad \text{independent of the oxygen pressure}$$

$$\text{or (b)}\quad H_2 + O_2 \rightarrow H_2O$$
$$CO + \tfrac{1}{2}O_2 \rightarrow CO_2$$

and the considerations already advanced support the second scheme.

Oxy-redox Levels

We may note that on taking the simplest assumption of water formation by reduction of a preformed oxide, we obtain for the rate of formation of water by reduction of the oxide

$$\frac{dp_{H_2O}}{dt} = K' p_{H_2} \times \theta$$

and for the formation of the oxide

$$-\tfrac{1}{2}\frac{dp_{O_2}}{dt} = K'' p_{O_2} \times (1-\theta)$$

where θ is the fraction of the surface covered by reducible oxide. We obtain

$$\frac{\theta}{(1-\theta)} = \frac{K'' p_{O_2}}{K' p_{H_2}}$$

This establishes the important point that the composition of the catalyst, viz. Cu^+/Cu or Cu^{2+}/Cu^+, is governed by the reactivity and the composition of the gas phase p_{O_2}/p_{H_2}. The oxy-redox potential (and thus the catalytic activity and the height of the Fermi level) is not a unique property of the catalyst *per se* but is dependent, when in the steady state, on the composition of the reacting gases.

A good visual example of this fact is seen on exposure of zinc chromite to carbon monoxide–oxygen mixtures where, even in 90% CO, the chromite is still in the coloured oxidized form.

Although the mechanism of water formation on a number of oxides is similar to that on copper, on metals having *d*-band vacancies such as platinum or palladium, chemisorption of hydrogen can take place, likewise by *d-s* promotion on silver and at high temperatures chemisorption may occur of hydrogen removable by gaseous oxygen. The chemisorbed hydrogen may be formed by heterolytic or homolytic splitting

$$Pd + H_2 \rightarrow Pd^+H + H^+ + 2e$$

or

$$2Ag + H_2 \rightarrow 2Ag^+H + 2e$$

Langmuir showed that platinum catalysts suffered from lack of good reproducibility; at high temperatures the rate of reaction $\omega \propto p_{O_2}/p_{H_2}$, but under certain conditions it appeared that the reaction was supplemented by attack of chemisorbed hydrogen by gaseous oxygen. At high temperatures not only was PtO_2 formed but there was additional oxygen loss due to condensation of atomic oxygen or ozone on the walls of the vessel. At low temperatures $\omega \propto p_{H_2}$ and it is significant that chemisorbed oxygen is readily removed by hydrogen and vice versa. On silver the rate was independent of the oxygen but varied approx. as $\sqrt{p_{H_2}}$. Of the various forms of oxygen present on a catalytic oxide surface, it is important to note the presence of physically adsorbed mobile molecular oxygen as well as lattice oxygen capable of exchange with surface chemisorbed oxygen which must be present in the ionic or polar form MO^-. Whereas the experimental data reveal that dissociative adsorption takes place, the phenomenon of coupled oxidation suggests that, at any rate at low temperatures, there is a temporary intermediary peroxide demanding the following sequence of steps on chemisorption

$$M + O_2 \rightarrow MO_{2\,phys} \rightarrow MO_2^- \rightarrow MO_2^{2-} \rightarrow 2MO^-$$

The presence of a molecular ionic form is likewise supported by spin resonance experiments.

The Oxidation of Carbon Monoxide

The oxidation of carbon monoxide has been the subject of many investigations. Not only does it oxidize on a variety of catalytic surfaces, but the temperature range is a wide one, extending from the Bonecourt boiler combustion to room temperature oxidation on silver–manganese dioxide mixtures such as hopcalite.

Many of the experimental findings are at variance with one another with respect to the order of the reaction. The data have been recorded in Table XXXII. Dilution with carbon dioxide has in general a negligible

TABLE XXXII

Catalyst	Order
Quartz glass	$p_{O_2}p_{CO}^{-1}$
Crystalline quartz	$p_{O_2}p_{CO}$
NiO room temp.	Zero
NiO below 170°C	$\sqrt{p_{O_2}p_{CO}}$ ($\Delta E = 2{\cdot}21$ kcal mole^{-1})
NiO above 250°C	p_{CO}
CuO	p_{CO}
ZnO	$\sqrt{p_{O_2}}$ Inhibition by great excess CO, acceleration by great excess O_2 low temperatures $\propto$ partial pressure of each gas
Pd, Pt	$p_{O_2}p_{CO}^{-1}$ High temperatures practically temperature independent
Ag at 100°C	p_{CO} excess $O_2 \propto p_{CO}$ excess CO $\propto p_{O_2}$
Ag above 100°C	$p_{CO}p_{O_2}$

effect but on thoria a slight inhibition has been observed. Evidently the behaviour of carbon monoxide is more complex than that of hydrogen. The decomposition of carbon monoxide by the Boudouard reaction takes place readily on metallic nickel and is frequently a complicating factor

$$2CO \rightarrow C + CO_2$$

On metals possessing *d*-band vacancies we obtain adsorption of CO. Two configurations have been detected by infrared analysis. These have been assigned to structures

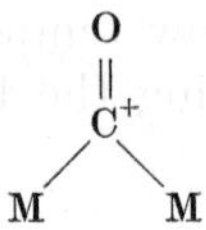

predominant on the [100] plane of nickel and O═C═M on the [110] plane. On tungsten some gas is relatively weakly held $\Delta H \sim 20$ kcal mole^{-1} falling to 12 on increasing coverage. The gas held on two sites is more strongly held having $\Delta H \sim 50$, 75, 100 kcal mole^{-1} depending on the lattic spacings on the different crystal planes. On nickel containing oxygen and on p-type semiconducting oxide catalysts the most active are those possessing a lattice spacing in the range 1·75–1·84 Å. An additional form is also believed to be present which has been given a configuration.

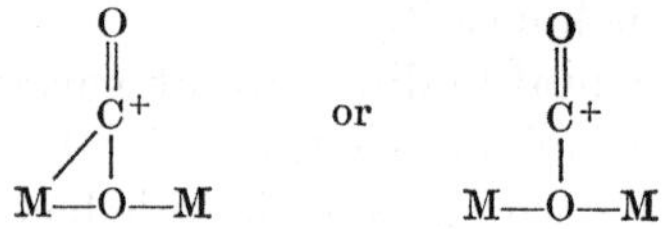

The carbon monoxide may be present as a carbonyl group on nickel metal or as a ligand in a crystal lattice of a cation such as Ni^{2+}. On oxidation of adsorbed carbon monoxide we should obtain a bidentate complex M—O—O + CO → MO—OCO which must eventually yield CO_2, possibly through the intermediary of a relatively stable carbonate ion. That, like hydrogen, reaction of carbon monoxide with chemisorbed oxygen is one of the reactions involved is revealed by consideration of the results on semiconductors. Zinc oxide is a good catalyst, at least after heating in carbon monoxide, and it is claimed that p-type semiconductors are more active than the n-type. Measurement of the temperature coefficient of the electrical conductivity of some oxides suggests that there is an optimum value of $E = 0{\cdot}4$ eV for CO oxidation. Values obtained for other oxide catalysts were: MnO_2, $E = 0{\cdot}72$ eV; Co_3O_4,

0·54 eV; CoO, $E = 0{\cdot}22$ eV. It may be noted that MnO_2 is usually n-type whereas there may be some doubt as to the respective activities of the two oxides of cobalt since Co_3O_4 when deposited on γ-Al_2O_3 is slowly converted into $CoAl_2O_4$ and the activity can be assigned to CoO.

In the crystal close-packed cubic form of NiO the octahedral sites are occupied by Ni^{2+} which, by increasing the oxygen content, can be partly converted into Ni^{3+}. These render the catalyst ferromagnetic and it is claimed that the energy of activation for the oxidation rises from some 3 to 15 kcal mole^{-1}. These changes in valency can be followed by change in colour and the p,n-type characteristics of NiO are very susceptible to admission of either CO or O_2, the former gas being capable of reducing the oxide to the metal and making d-band electrons available. Thus, as in the case of copper in the oxidation of hydrogen, we can have, dependent on the partial pressures of the two gases and the temperature, a catalyst composed of the oxy-redox systems which may involve

$$\frac{Ni^{3+}}{Ni^{2+}}, \frac{Ni^{2+}}{Ni^{+}}, \frac{Ni^{+}}{Ni^{0}}$$

Doping CoO with lithium renders it more active. Doping NiO with lithium likewise renders it more active, but alters its kinetic behaviour towards that obtaining at low temperatures.

We may note also that CO adsorbed on thoria, nickel oxide and zinc oxide can be removed by means of oxygen as well as the reverse process. We have already seen (p. 36) from the Hall effect coefficient that chemisorbed oxygen can react readily with CO.

On exposure to oxygen the electron concentration of zinc oxide and doped oxides according to Chon and Prater falls at a rate equal to the rate of oxygen uptake, i.e. it follows the Elovitch equation for the rate of a dissociative adsorption. The rate of electron loss corresponded to the reaction $O_a \rightarrow O_a^{-1}$. On admission of carbon monoxide the electron concentration rises in amount corresponding to the reaction $CO + O_a^{-1} \rightarrow CO.O + e$. Electron transfer does not take place when CO is adsorbed on zinc oxide in the absence of chemisorbed oxygen, although CO is an electron donating molecule.

The Oxidation of Sulphur Dioxide

The catalytic oxidation of sulphur dioxide to the trioxide over platinum is historically important since it formed the basis of Bodenstein's diffusion layer hypothesis of the mechanism of the catalytic process. This hypothesis represents a restatement of the ideas advanced by Faraday.

The experiments of Bodenstein and Finck revealed that the rate of oxidation of sulphur dioxide was a function of the partial pressures of the reactants and of the product such that

$$\omega = \frac{Kp_{SO_2}}{\sqrt{p_{SO_3}}}$$

at high pressures of oxygen and

$$\omega = \frac{Kp_{O_2}}{\sqrt{p_{SO_3}}}$$

at high pressures of sulphur dioxide. It was imagined that the rate of reaction was governed by the rate of diffusion of the low-pressure reactant through a layer of absorbed product sulphur trioxide, the thickness of which varied as the square root of its partial pressure.

Although diffusive limitations to the rate of reaction at the surface are to be met in cases such as in combustion or in hydrogenation in liquid media, it seems improbable that such is the case here. Langmuir pointed out that if the sulphur trioxide is strongly adsorbed on two elementary spaces and the surface remains practically completely covered with this gas we may write

$$\alpha p_{SO_3}\theta^2 = \nu$$

where θ is the fraction of the surface bare, αp_{SO_3} and ν the rates of condensation and evaporation of the trioxide. The rate at which SO_2 or oxygen can enter into reaction at the surface will be proportional to $p_{SO_2}\theta$ or $p_{O_2}\theta$. Hence

$$\omega \propto \frac{p_{SO_2}\sqrt{\nu}}{\sqrt{\alpha p_{SO_3}}} \quad \text{or} \quad \frac{p_{O_2}\sqrt{\nu}}{\sqrt{\alpha p_{SO_3}}}$$

depending on which gas is in excess.

We have already noted in the cases of the catalytic oxidation of hydrogen and of carbon monoxide that electron removal from and return to the substrate is involved in the process of alternate oxidation and reduction. This is reflected in a change in valency of an electron donating ion. In the oxidation of sulphur dioxide, vanadium pentoxide is an effective catalyst and, as in the cases already considered, we might anticipate that in the steady state the rate of oxidation of the reduced form would be equal to the rate of reduction of the oxidized form or

$$\alpha p_{SO_2} V^{5+} = \beta p_{O_2} V^{4+}$$

or

$$\frac{V^{5+}}{V^{4+}} = \frac{\beta p_{O_2}}{\alpha p_{SO_2}}$$

The composition of the catalyst will thus depend on the partial pressures and reactivities of the sulphur dioxide and oxygen. In practice, the optimum composition for a stoichiometric gas mixture is said to correspond to V_4O_3 or $V^{5+}/V^{4+} = 1{\cdot}17$ and, since two moles of SO_2 are oxidized per mole of oxygen, it is evident that the oxidation of the SO_2 proceeds more readily than the oxidation of the V^{4+}.

Influence of Crystal Field

As in other oxides, the energetics of the cation oxygen ligand reaction are mainly determined by the crystal field. There are a number of cases in which the transition structure of a metallic ion in its crystal matrix have been established. These include V^{5+}, Cr^{5+} and the hydrate of nickel sulphate.

The lattice structure of V_2O_5 consists virtually of distorted octahedral VO_6 units in which the V—O distances are 1·88, 1·88, 1·77 and 2·02 Å together with two ligands one of which corresponding to V═O is very short, viz. 1·54 Å stabilized by π-bonding between filled oxygen p-orbitals and empty metallic ion d-orbitals. The other ligand opposite to this is much longer, viz. 2·83 Å and it is this oxygen which is labile and takes part in the catalytic process. On removal of the oxygen the lattice structure still remains characteristic of V^{5+} and does not change to the V^{4+} structure which is of the rutile type, i.e. the V^{4+} left on removal of the oxygen must be regarded as a pseudomorph. The strength of this V—O bond is most readily altered by changing the environment of the double bonded V═O. This can be detected by infrared analysis since the V—O stretching frequency $\nu = 1025\ cm^{-1}$ is shifted to longer wavelengths by promoters such as alkaline sulphates, where the order of activity in promotion is $Cs > Rb > K > Na$. This enhancement of catalytic activity by weakening the bond strength V═O is also observed in the oxidation of carbon monoxide as well as in the case of sulphur dioxide. The mechanism can also be regarded as being governed by the chemisorption of oxygen on to an n-type semiconductor. A vacant coordination site is also present in Cr^{5+} and for it to react with a hydrocarbon to form the transition structure, the metal ion orbital and that of the hydrocarbon must be compatible and overlap. The addition of R to an olefin may be depicted as follows.

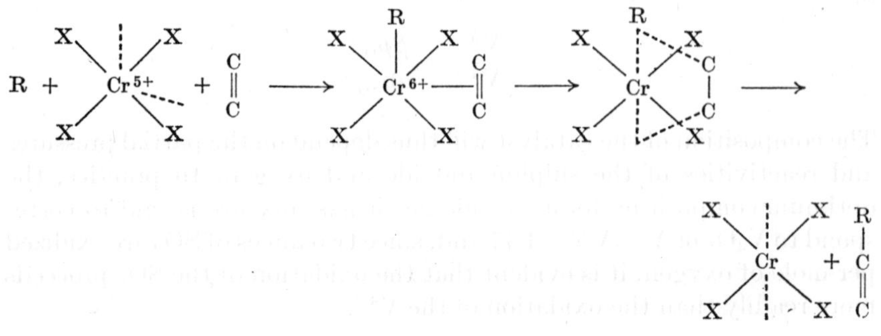

The catalytic activity of nickel sulphate in several cases of polymerization is a function of its acidity and this is due to the presence of Lewis sites. Nickel sulphate hydrate is the most active form and it has been claimed that Ni^{5+} has a vacant sp^3d^2-orbital which gives it the Lewis acid properties.

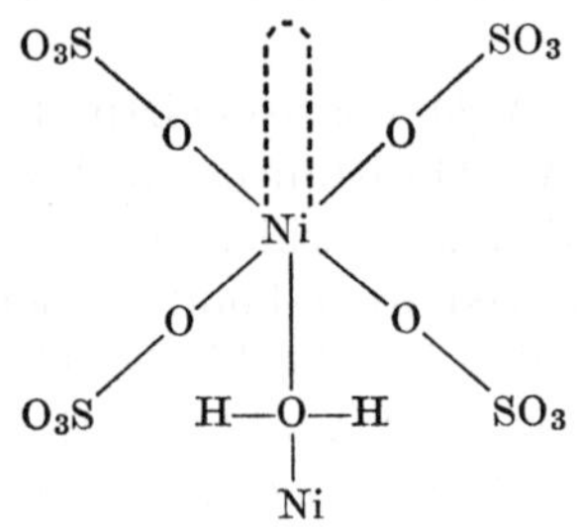

The Kinetics of Oxidation Catalysis

We have noted that there is a wide variation in the experimentally determined reaction kinetics of the oxidation of hydrogen, carbon monoxide and sulphur dioxide, depending on the nature of the catalyst, the temperature and composition of the reacting gases. Adopting the premise that the mechanism involves the alternate oxidation and reduction of the substrate followed by desorption of the products, various kinetic schemes can be advanced taking into consideration the relative importance of the individual steps in the complete operation. Amongst the many possibilities, three appear to be of importance differing with respect to the increasing adsorption of a reactant or product. These can be formulated in the following manner, taking the oxidation of carbon monoxide by a catalyst as the example.

Case I	Case II	Case III
$M+O_2 \rightleftharpoons 2MO$	$M+O_2 \rightarrow 2MO$	$M+O_2 \rightarrow 2MO$
$MO+CO \rightarrow MO.CO$	$M+CO \rightleftharpoons MCO$	$M+CO \rightarrow MCO$
$MO.CO \rightleftharpoons M+CO_2$	$MO+CO \rightarrow MO.CO$	$MO+CO \rightarrow MO.CO$
	$MOCO \rightarrow M+CO_2$	$M.CO+O_2 \rightarrow MO.CO$
		$MO.CO \rightleftharpoons M+CO_2$

We note that the oxygen is more strongly adsorbed than the carbon monoxide and that there is an increase in the irreversibility of the individual steps as we proceed from case I to case III where the adsorption of the oxygen and that of the carbon monoxide are irreversible.

In case I the oxygen is regarded as reversibly adsorbed as well as the product, the reaction taking place between chemisorbed oxygen and gaseous or physically adsorbed carbon monoxide. In case II the same reaction process is involved, but the oxygen is regarded as irreversibly adsorbed, the carbon monoxide adsorption being reversible. In case III there are two reactions, one between gaseous carbon monoxide and chemisorbed oxygen, and one between chemisorbed carbon monoxide and gaseous oxygen. The slow step in the latter case is the evaporation of the product carbon dioxide or the conversion of the reaction complex MO.CO and its subsequent evaporation as carbon dioxide and reformation of a reaction site.

Case I

For case I we can adopt Langmuir's method to investigate the dependence of the reaction velocity on the partial pressures of reactant and product and their lifetimes or adsorbabilities when dynamic equilibrium is established. We consider a fraction $1-\theta_1-\theta_2$ of the surface to be bare, fractions θ_1 and θ_2 being covered with chemisorbed oxygen and carbon dioxide respectively.

The conditions for equilibrium between gaseous and chemisorbed oxygen are given by (4).

$$K\sqrt{p_{O_2}}(1-\theta_1-\theta_2) = \nu_1\theta_1 \quad \text{or} \quad \nu_1\theta_1 = \frac{K\sqrt{p_{O_2}}}{1+K\sqrt{p_{O_2}}+K'\sqrt{p_{CO_2}}} \tag{4}$$

where $K\sqrt{p_{O_2}}$ is the rate of chemisorption of oxygen and ν_1 is its specific rate of evaporation.

Similarly (2), for the carbon dioxide equilibrium

$$K'\sqrt{p_{CO_2}}(1-\theta_1-\theta_2) = \nu_2\theta_2 \quad \text{or} \quad \nu_2\theta_2 = \frac{K'\sqrt{p_{CO_2}}}{1+K\sqrt{p_{O_2}}+K'\sqrt{p_{CO_2}}} \tag{5}$$

The reaction takes place (6) between chemisorbed oxygen and gaseous carbon monoxide or

$$\omega = Kp_{CO}\theta_1 = \frac{K_k p_{CO}\sqrt{p_{O_2}}}{1+K\sqrt{p_{O_2}}+K'\sqrt{p_{CO_2}}} \tag{6}$$

where ω is the rate of reaction. The limiting cases when θ_1 or $\theta_2 \sim 1$ indicate that when there is but little adsorption, i.e. θ_1 and θ_2 small, at high temperatures $\omega \propto p_{CO}\sqrt{p_{O_2}}$. When the oxygen is strongly adsorbed $\omega \propto p_{CO}$, and when the product CO_2 is strongly adsorbed $\omega \propto p_{CO}/\sqrt{p_{CO_2}}$. If we consider the adsorption of the carbon monoxide, i.e. θ_3 being part of the surface, we find that in excess of carbon monoxide $\omega \propto \sqrt{p_{O_2}}$. If the reaction takes place between physically adsorbed carbon monoxide and chemisorbed oxygen and the former obeys the Langmuir isotherm, we obtain

$$\omega = \frac{ap_{CO}}{(1+bp_{CO})}\frac{\sqrt{p_{O_2}}}{(1+K\sqrt{p_{O_2}}+K'\sqrt{p_{CO_2}})}$$

These conditions appear to hold for the oxidation on a number of catalytic oxides.

Case II

We can, in a similar manner, derive expressions for the reaction kinetics for case II. Here the oxygen is considered to be irreversibly chemisorbed, e.g. by a metal. The rate of adsorption is consequently a measure of the rate of reaction with the carbon monoxide in the gas phase, chemisorbed carbon monoxide being inactive. The products are assumed not to inhibit the process, but to evaporate or be readily displaceable. These conditions are fulfilled by the following relationships.

The irreversible adsorption of oxygen is equal to the rate of reaction

$$\omega = kp_{O_2}\theta \tag{7}$$

The carbon monoxide is in reversible equilibrium with the surface phases.

$$K'p_{CO}\theta = \nu\theta_2 \tag{8}$$

where ν is the specific rate of evaporation of the monoxide. The reaction takes place between chemisorbed oxygen and gaseous carbon monoxide

$$\omega = kp_{CO}\theta_1 \tag{9}$$

The reaction rate involving the relationship $\theta+\theta_1+\theta_2=1$ is given by

$$\frac{\omega}{Kp_{O_2}}+\frac{\omega}{K'p_{CO}}+\frac{\omega}{kp_{O_2}}\times\frac{Kp_{CO}}{\nu} = 1$$

When the limiting values of θ, θ_1 or θ_2 attain unity for $\theta \sim 1$ there is but little adsorption as at high temperature $\omega \propto p_{O_2}$, when the oxygen is strongly adsorbed, i.e. $\theta_1 \sim 1$ $\omega \propto p_{CO}$ and when the carbon monoxide is strongly adsorbed, i.e. $\theta_2 \sim 1$ we find

$$\omega \propto \frac{Kp_{O_2}}{K'p_{CO}}$$

Langmuir, in his classical paper on the catalysis of the oxidation of carbon monoxide and hydrogen at a platinum surface found all these conditions to hold.

Case III

We have seen in case I that only the oxygen is chemisorbed and that reversibly; in case II the adsorption is stronger so that the oxygen is chemisorbed irreversibly and the carbon monoxide chemisorbed, but in a reversible manner. In case III both the oxygen and the carbon monoxide are assumed to be irreversibly adsorbed and the reaction is consequently governed by the evaporation of the product.

Dynamic equilibrium is attained where the surface is almost completely covered by the product CO_2 or the complex MO.CO. Equilibrium between the rate of evaporation and condensation is given by the relationship

$$\nu_{CO_2} = K\sqrt{p_{CO_2}}\,\theta$$

The fraction θ is small and can be regarded as the vacant sites for which oxygen, carbon monoxide or carbon dioxide can compete, the two former being held irreversibly and only removable by chemical reaction with a component from the gas phase.

The rate of filling the sites with oxygen and carbon monoxide is given by $K'p_{O_2}\theta$ and $Kp_{CO}\theta$ respectively to give θ_1 and θ_2. The sites so filled are then emptied again by reaction with carbon monoxide and oxygen from the gas phase. These rates are $K'p_{CO}\theta_1$ and $Kp_{O_2}\theta_2$. At equilibrium the rates of filling and emptying must be equal or

$$K'p_{CO}\,\theta = Kp_{O_2}\,\theta_2 \quad \text{and} \quad Kp_{O_2}\,\theta = K'p_{CO}\,\theta_1$$

The rate of reaction, being the sum of the reaction velocities on the sites, is accordingly

$$\omega = K'p_{CO}\,\theta_1+Kp_{O_2}\,\theta_2 = (Kp_{O_2}+K'p_{CO})\,\theta$$

Hence

$$\theta = \frac{\omega}{Kp_{O_2}+K'p_{CO}} = \frac{\nu_{CO_2}}{K\sqrt{p_{CO_2}}}$$

or

$$\omega = \frac{\nu_{CO_2}(Kp_{O_2}+K'p_{CO})}{K\sqrt{p_{CO_2}}}$$

The reaction rate is consequently determined by the sum of the partial pressures of the reactants, and if the partial pressure of one reactant remains constant the rate rises on the increase of the partial pressure of the other.

These conditions are found to hold in the oxidation of hydrogen at a copper surface to form water and of sulphur dioxide at a platinum surface to form the trioxide. It is of interest to observe that Langmuir noted that on active platinum the low-temperature oxidation appeared to involve the oxidation of chemisorbed hydrogen as well as the inverse process.

On strongly p-type semiconductors we might anticipate that carbon monoxide, an electron donor, might be more strongly adsorbed than oxygen. This appears to be the case for thoria and for thoria doped with ceria. Oxygen is hardly or very weakly adsorbed, but carbon monoxide more strongly. On admission of oxygen to a thoria surface covered with carbon monoxide, the gas is recovered as carbon dioxide. We have, in consequence, the reaction steps

$$\begin{aligned} M+CO &\rightleftharpoons M.CO \\ M.CO+\tfrac{1}{2}O_2 &\rightarrow MOCO \\ MOCO &\rightleftharpoons M+CO_2 \end{aligned}$$

which can be contrasted with case I for n-type semiconductors and leads to the kinetic expression for the reaction velocity.

$$\omega = \frac{K_n\sqrt{p_{O_2}p_{CO}}}{1+K'p_{CO}+K'p_{CO_2}}$$

In all these cases of oxidation we note that the experimental findings support the view that the reactions involve a chemisorbed and a gaseous reactant and such factors as interaction between two chemisorbed reactants or surface migration need not be considered. Further, whereas the most common catalysts involve the oxygen as the chemisorbed reactant, other catalysts can chemisorb both reactants, each attacked from the gas phase, and a few catalysts involve adsorption of the oxidizable reactant molecule and a change from an n- to a p-type semiconductor.

The Oxidation of Ethylene

In the oxidation of olefins, ethylene is unique in that large conversions to ethylene oxide can be obtained by catalytic oxidation at a silver surface. In the oxidation process carbon dioxide and water are produced simultaneously, but these products are the result of two independent reactions: the direct oxidation of ethylene and the further oxidation of ethylene oxide in some isomeric form. Oxygen is rapidly chemisorbed by silver to depths exceeding a monolayer. Isotopic oxygen $^{16}O^{18}O$ experiments reveal that the oxygen is freely exchangeable with gaseous oxygen. The value of E_{ad} varies from 12·7 to 29 kcal mole^{-1}, whereas the heat of adsorption varies from $\Delta H_{ad} = 29$ to 12·7 kcal mole^{-1}. The activation energy for desorption is $E_{des} = 33$ kcal mole^{-1}, which also governs the rate of isotopic exchange. Since penetration in depth occurs, the adions must be mobile. The face-centred silver lattice has spacings 2·88 and 4·08 Å, the former can retain an oxygen adion whereas the latter permits the formation of a pair of adions in which the separation is close to that in the oxygen molecule, and it is possible that exchange takes place at these centres.

Twigg found that the contact resistance of silver is greatly increased on the adsorption of oxygen, ethylene oxide and acetaldehyde, and is unaffected by ethylene; thus the primary action is the chemisorption of oxygen,

$$O_2 + Ag \rightarrow Ag^+O^-(ads)$$

which can be removed by extraction with ethylene (Rideal mechanism):

$$Ag^+O^-(ads) + C_2H_4 \rightarrow C_2H_4O + Ag$$

The passage of ethylene oxide, in nitrogen as a carrier, reveals by Schiff's reagent that the ethylene oxide is in part converted into its isomer acetaldehyde which, again in the presence of oxygen, is rapidly oxidized to carbon dioxide and water. The rate of this oxidative process is some hundred times greater than that of the original oxidation to ethylene oxide, and acetaldehyde is not, consequently, detectable in the gas stream of the products; thus, the rate of production of carbon dioxide from this source depends on the rate of isomerism of the ethylene oxide to acetaldehyde.

$$AgC_2H_4O \rightarrow AgCH_3CHO$$
$$AgO + CH_3CHO \rightarrow CO_2 + H_2O$$

The total carbon dioxide production when ethylene–air mixtures are passed over a silver catalyst is in excess of that produced by the further oxidation of the ethylene oxide and arises from the direct oxidation of the ethylene

$$C_2H_4 + AgO(ads) \rightarrow CHO + CHO \rightarrow CO_2 + H_2O$$

The simplest interpretation of these results is that ethylene oxide results from the interaction of ethylene with a single chemisorbed oxygen adion, whereas the direct conversion of ethylene to carbon dioxide is brought about by interaction of ethylene with a pair of oxygen adions; from conductivity measurements on silver films the rate is indeed found to be proportional to the square of the concentration of oxygen adions. Ethylene oxide when passed over silver leaves an insoluble carbonaceous residue and the resistance of the film increases, indicating a reaction

$$C_2H_4O + Ag \rightarrow Ag^{+}O^{-} + (C_2H_4)_n$$

The general kinetic scheme can accordingly be written

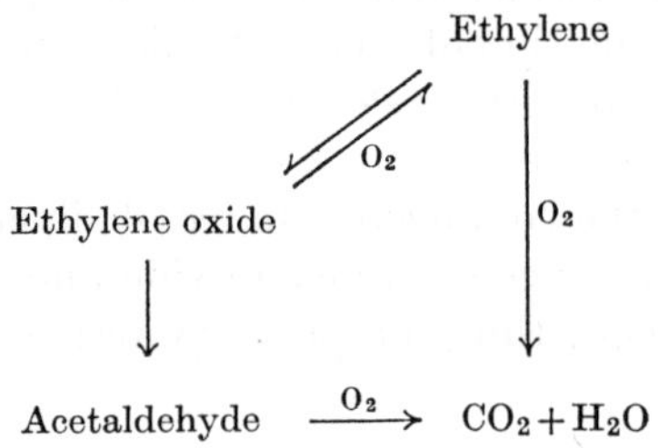

and the limitations in the choice of catalyst become evident from the energy diagram (Fig. 8).

The greater the heat of formation of the adsorbed reaction complex, i.e. the smaller ΔH, the smaller will be yield of ethylene oxide. Indeed, there is an upper limit above which ethylene oxide cannot be formed.

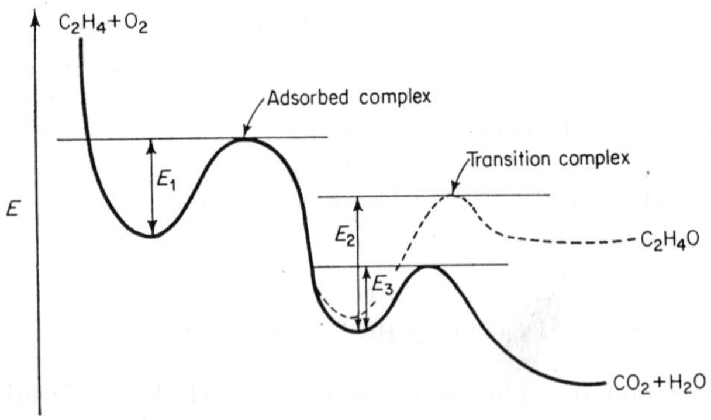

FIG. 8.

The relative magnitudes of E_1 and E_2, the energies of activation to form the transition complexes, depend not only on this adsorption heat but also on the pattern or structure of the transition complex which may be different for the two products.

Oxidation of Ammonia and Saturated Hydrocarbons

The oxidation of ammonia carried out with platinum or platinum–rhodium gauze as catalyst at 850°C results in its conversion to nitric oxide. The process is relatively efficient with a yield under optimum conditions of over 80% nitric oxide, the rest being lost as elementary nitrogen. Photomicrographs of the platinum gauze after it has served as an active catalyst reveal that there has been a great disturbance of the original crystal planes, the surface being covered with a "cauliflower" like excrescence. At some 1100°C platinum is directly attacked by oxygen to give the volatile oxide which, on analysis, is found to be PtO_2, and it would appear most likely that the oxidation of ammonia proceeds through the reduction of the surface oxide to form a surface imide

$$PtO + NH_3 \rightarrow Pt{=}NH + H_2O$$

The surface residence time of the imide being longer than that of either $Pt—NH_2$ or of an amine complex $Pt—NH_3$ favours this as the primary intermediary in the conversion to nitric oxide. The fact that hydrogen added to the gaseous ammonia cuts down the rate of thermal decomposition of ammonia on the catalyst supports the view that there must be present on the surface an intermediary, e.g. NH which can equilibriate with the gas $NH(ads) + H_2 \rightleftharpoons NH_3$.

The mechanism of the oxidation of hydrocarbons to form intermediary oxidation products, such as that of methane to formaldehyde or to a mixture of carbon monoxide and hydrogen, has not been elucidated. It is believed that hydroxylation processes play a major part in the complex reaction that takes place in oxidation on catalytic metals which readily form surface oxides, e.g.

$$CuO + CH_4 \rightarrow CuCH_3OH \rightarrow CuCH_2O$$

$$\text{Propene} + Cu^{2+} + O^- \rightleftharpoons Cu^+ + \text{acrolein}$$

Such measurements as are available indicate that active metal and metal–oxide systems are effective electron donors to oxygen and hydrocarbons when these are chemisorbed.

Methyl alcohol can be oxidized with good yield to formaldehyde over catalysts such as ferric molybdate. The reaction rate at approximately

270°C is first-order in respect to the methyl alcohol and zero-order in the oxygen. In its active state the molybdate catalyst is an n-type semiconductor that cuts down the adsorption of the methyl alcohol in the reducing atmosphere. Hydrocarbons such as tetralin and cyclohexene can readily be oxidized in the liquid state with the aid of solid chromic, cobalt or manganese oxides. The reaction rate is independent of the oxygen pressure provided that the solution is maintained in equilibrium with the gas phase, but it is proportional to the hydrocarbon concentration. Free radicals formed at the surface of the oxide initiate oxidation chain reactions but chain termination is homogeneous. Free radical initiation of reaction chains can of course be brought about with soluble organometallic catalysts such as cobalt acetate or stearate. The steps involving initiation, propagation and termination are

$$\begin{aligned} RH &\xrightarrow{C} R\ k_1 \text{ (initiation)} \\ R+O_2 &\longrightarrow RO_2 \\ RO_2+RH &\longrightarrow RO_2H+R\ k_2 \text{ (propagation)} \\ 2RO_2 &\longrightarrow \text{products } k_3 \text{ (termination)} \end{aligned}$$

The free radicals produced can act as initiators in radical polymerization: thus, from the radicals R and RO_2 we have two initiators

$$RO_2+\text{styrene} \rightleftharpoons \text{complex which can be broken down by hydrolysis}$$
$$R+\text{styrene} \rightleftharpoons \text{complex that is stable}$$

The ratio in which the two species of polymeric styrene chains are formed can be determined with the aid of isotopic carbon; e.g., for benzyl peroxide we can initiate with $C_6H_5{}^{14}COO$ or with $^{14}C_6H_5^+$. Initiation of chains can also be effected with •OH radicals provided by the catalytic decomposition of hydrogen peroxide, e.g.

$$Fe^{2+}+H_2O_2 \rightarrow [Fe^{2+}.OH.OH] \rightarrow [Fe^{3+}.OH]+\cdot OH$$
$$CH_2{=}CH_2+\cdot OH \rightarrow \cdot OH{-}CH_2{-}CH_2{-}$$

With other persalts both radical and ionic mechanisms may intrude, e.g.

$$S_2O_5^{2-}+R_r \rightarrow SO_4^-+SO_4^{2-}+R_r^+$$

where R_r is a reducing agent.

$$\begin{aligned} SO_4^-+\text{monomer} &\rightarrow [\text{monomer–}SO_4]^- \text{ complex} \\ SO_4^-+H_2O &\rightarrow HSO_4^-+OH \\ OH+\text{monomer} &\rightarrow [\text{OH–monomer}] \text{ complex} \end{aligned}$$

Whereas the catalytic oxidation of ethylene to ethylene oxide on a silver catalyst involves chemisorbed oxygen being attacked by ethylene, the catalytic oxidation of other olefins namely of propylene to acrolein and of butylene to butadiene follows a different mechanism. The most satisfactory catalyst is a bismuth molybdate at about 450°C.

It has been shown by means of ^{14}C and deuterium that in the oxidation of propylene an allyl intermediary is formed which is rapidly oxidized further to the aldehyde,

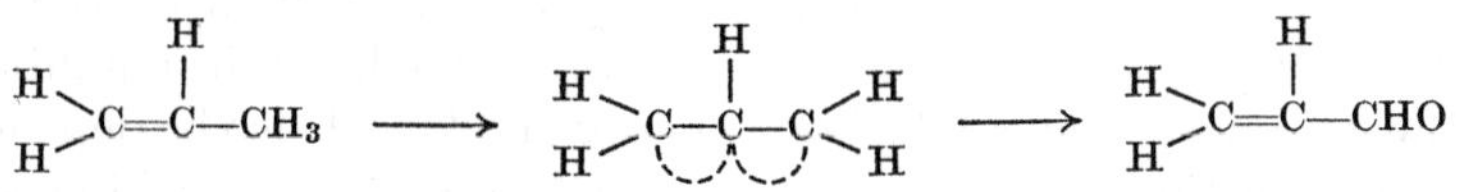

since the two carbon atoms were found to be identical and no rapid isomerization of the propylene took place on this catalyst under these oxidation conditions,

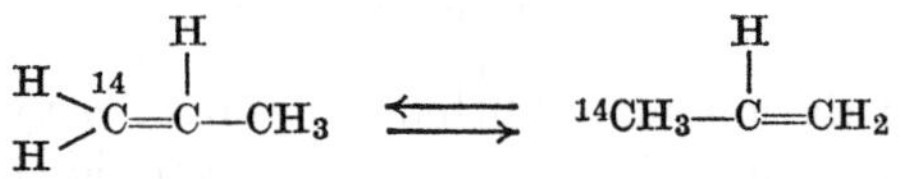

The allyl intermediary may be the result of the catalyst acting as a carbonium ion formation system

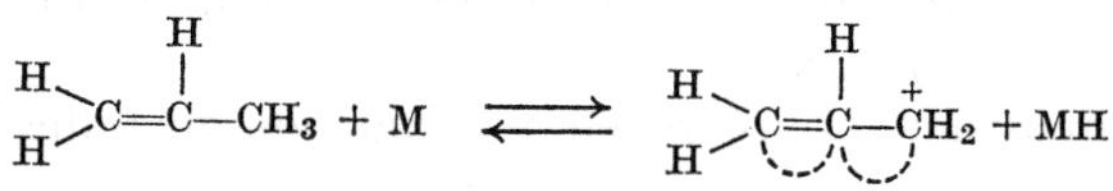

but hydrogen abstraction to form a radical also can take place by reaction with chemisorbed oxygen

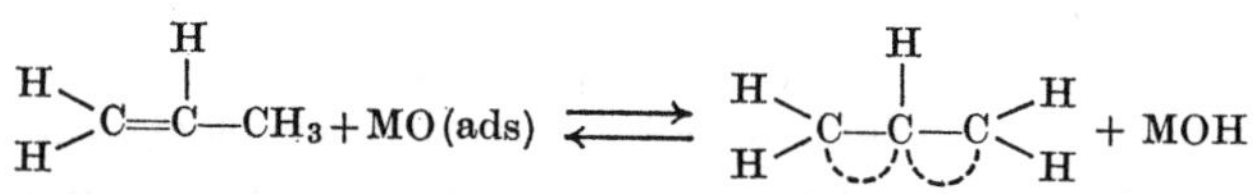

followed by the further abstraction

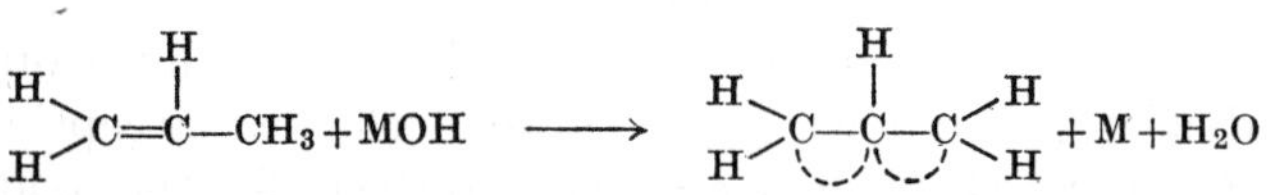

and reoxidation of the catalyst

$$2M + O_2 \rightarrow 2MO(ads)$$

The oxidation rate is found to be proportional to the olefin partial pressure and independent of the oxygen or of the adsorbed products, although with C_5 and higher olefins the kinetics have been given as

$$\omega = kp_{\text{olefin}}\left(\frac{\alpha p_{O_2}^{1/2}}{1+\alpha p_{O_2}^{1/2}+\beta p_{\text{olefin}}}\right)$$

The rate-controlling mechanism would be the dissociative adsorption of the olefin if carbonium ions are formed or a Rideal mechanism reaction of the gaseous olefin with chemisorbed or oxidized catalyst if an extractive mechanism is envisaged; this latter mechanism appears to be the more probable.

It is worth noting that the p- or n-type characteristics of the bismuth molybdate or its oxy-redox potential would be dependent on the ratio M/MH or MO/M in the surface phase which should be in equilibrium with the gas phase. Hence, the bond energy M—O will also be dependent on the composition of the gas phase and this bond energy will be some measure of the energy of activation for the reaction.

We thus obtain

$$\frac{M^{4+}}{M^{6+}} \rightarrow \frac{[\text{aldehyde}][H_2O]}{[\text{hydrocarbon}][O_2]}$$

The gas-phase oxidation of naphthalene and of *o*-xylene to phthalic anhydride proceeds smoothly on vanadium pentoxide catalysts yielding some 86·5% yield in the case of naphthalene but only 70% for *o*-xylene. This is due to the formation of *o*-toluic acid. Increase in the reactivity of the second methyl group can be brought about by the addition of small quantities of halogens.

The decomposition of hydrogen peroxide on metallic oxides and spinels may be regarded as fundamentally an electron accepting process.

$$\begin{array}{ccc} H_2O_2 & & OH^- + OH \\ e\uparrow & \rightarrow & \\ Fe^{2+} & & Fe^{3+} \end{array}$$

We thus note that with such metallic oxides two electronic processes can take place. We may have lattice defects in the system, and the electron transfers can be interpreted on the basis of p- or n-type semiconductors acting as sinks or sources of electrons to build up a double layer; in addition, the oxides of the transition metals must often have unpaired *d*-electrons and we get a Dowden electron-accepting mechanism. Cr_2O_3 or Fe_2O_3 are not effective catalysts for the oxidation of carbon monoxide or for N_2O decomposition, although they have unpaired *d*-electrons, yet they are placed correctly from their semiconductivity point of view (n-type).

Chapter 6

Catalytic Hydrogenation

The versatility of catalytic hydrogenation has been recognized since the classic work of Sabatier. It forms the basis of a number of important industrial processes such as the methanation of carbon monoxide, the synthesis of ammonia and the saturation of unsaturated fats and waxes. It finds increasing application in preparative organic chemistry, and the development of the petrochemical industry has been greatly facilitated by it.

As in the case of oxygen, our understanding of the behaviour of hydrogen with catalytic surfaces has been materially assisted by using the isotopic forms deuterium and, more recently, tritium. Further information has been gained from investigations on the ortho–para conversion of both hydrogen and deuterium.

The Dissociation of Hydrogen

Langmuir showed that metal wires at high temperatures could act as catalysts for the thermal dissociation and recombination of hydrogen.

$$H_2 \rightleftharpoons 2H$$

The determination of the rate of dissociation involves the removal of the atomic species and ensuring its absence in the gas phase. Getters such as molybdenum oxide or potassium are more effective than glass for this purpose. At high temperatures there are two reactions which by the principle of microscopic reversibility and detailed balancing are possible: thus, a molecule may strike the surface where there are two juxtapositioned vacant sites and be adsorbed on them as atoms, these may evaporate singly or, subsequently, two neighbouring atoms may combine and evaporate as a molecule; alternatively, a molecule can hit a vacant site and leave an adsorbed atom, the other atom returning to the gas phase, and the adsorbed atom may subsequently evaporate. Taking the first and simpler case, if the condensation coefficient is α

and $1-\theta$ is the fraction of the surface which is bare, the rate of condensation of molecular hydrogen will be $kP\alpha(1-\theta)^2$ and the rate of evaporation of hydrogen as molecules will be $\nu_{H_2}k'\theta^2$, where ν_{H_2} is the specific rate of evaporation or the reciprocal of the lifetime of the molecular species. If this molecular process is in dynamic equilibrium:

$$\nu_{H_2}k'\theta^2 = kP\alpha(1-\theta)^2$$

and the rate of atom formation, if sufficiently slow not to affect the molecular process, will be $\nu_H k''\theta$, where ν_H is the specific rate of evaporation of atomic hydrogen. The rate of production of atoms as a function of the pressure is

$$\omega = \nu_H k''\theta = \frac{k''\nu_H\left(\dfrac{kP\alpha}{k'\nu_{H_2}}\right)^{1/2}}{1+\left(\dfrac{kP\alpha}{k'\nu_{H_2}}\right)^{1/2}}$$

For the alternative mechanism the rate of production of atoms is clearly $\omega = 2k'''P\alpha(1-\theta)$ which when $\theta \sim 1$ approximates to $k''''(P)^{\frac{1}{2}}$.

Experimentally, the data indicate that the first mechanism is probably operative. Thus at low pressures the rate of atom production is proportional to the square root of the hydrogen pressure and becomes independent of the pressure when the surface is covered. This change has been noted to occur at about 10^{-5} mm pressure at 1700°K on platinum, and 1800°K on tungsten.

Values for ω the rate of production of atoms on various metals are:

$$(1{\cdot}36 \pm 0{\cdot}6)\times 10^{25}\sqrt{P}\,e^{(-51{,}000\pm 1200)/RT} \text{ on platinum}$$

$$(0{\cdot}23 \pm 0{\cdot}17)\times 10^{25}\sqrt{P}\,e^{(-51{,}000\pm 1600)/RT} \text{ on gold;}$$

$$(1{\cdot}5 \pm 0{\cdot}5)\times 10^{25}\sqrt{P}\,e^{(-52{,}000\pm 1000)/RT} \text{ on tungsten;}$$

P is expressed in mm Hg. We may observe that at very low pressures the rate of production of atoms may become proportional to the pressure and the second mechanism may also contribute to the production of atoms.

The Adsorption of Hydrogen

Investigations at lower temperatures important in catalytic and exchange reactions reveal that the system is complex. Methods of investigation include measurement of the "sticking coefficients" and rates of desorption as functions of the temperature and pressure, examination of the changes in electrical resistance and surface potential of the adsorbing metal and the calorimetric determination of the heats of adsorption as well as visual observation by field emission. Tungsten and nickel have been the metals most subject to investigation.

At low temperatures, below 100°K, the heat of adsorption is, except for a small initial portion, $\sim$1–2%, with a high heat of adsorption, independent of the surface coverage. Above this temperature the heat of adsorption falls with increasing coverage. It has been claimed that the integral heats of adsorption both for the constant as well as for the falling heat curves are the same, but the isosteric heats of adsorption are probably somewhat greater at the higher temperatures.

The low-temperature constant heat adsorption can be interpreted either on the basis that the condensation coefficient is the same for all parts of the surface and the subsequently adsorbed gas is not mobile or, alternatively, the gas physically adsorbed in the molecular or molecular ionic form is mobile over the surface whether the surface is bare or not and that its conversion to the chemisorbed atomic form is confined to the interface between the chemisorbed phase and the bare metal. This interfacial growth commencing at nuclei formed at high energy sites of surface dislocations will proceed with a constant linear heat evolution.

As the temperature is raised the gas becomes mobile and sites of higher energy are preferentially occupied. The change from an immobile to a mobile adsorption may take place over a temperature range. Sweet and Rideal concluded that the fraction of the adatoms which were mobile at any one temperature increased with the coverage. They give the data in Table XXXIII for the adsorption of hydrogen by nickel at 25°C.

Measurements of the rate of adsorption and of desorption reveal discontinuities both in the condensation coefficients and in the evaporation rates. The condensation coefficients are found to be constant or to

TABLE XXXIII

Fractional coverage θ	Fraction of adsorbed H migrating	Isosteric heat (kcal mole^{-1})	Activation energy for surface diffusion (kcal mole^{-1})
0·4	0·015	20·6	25
0·5	0·064	17·4	1·6
0·6	0·17	13·4	1·0
0·7	0·34	9·6	0·63
0·8	0·56	6·0	0·34
0·9	(0·88)	3·9	0·076

have slight negative temperature coefficients: values ranging from 0·135 to 0·020 have been recorded. Evaporation experiments suggest that the rate varies not only with the temperature at which the gas has been adsorbed but, if the rates are regarded as the rates of decomposition of discrete surface compounds, there must be at least three and possibly as many as six such structures operative in decomposition in the regions 77–190°K, 190–300°K, 300–600°K respectively. These give rise to discontinuities in both the adsorption–pressure and the adsorption–temperature curves. These three forms of adsorbed hydrogen have been termed the β-, α- and γ-forms in order of decreasing stability. The γ-phase is not found on adsorption above some 100°K but the α-phase is always present. The heat of formation of the γ-phase is thus low, being of the order of 6–10 kcal mole^{-1}. On desorbing gas adsorbed below 100°K, the γ-phase is found to be present but its presence can be detected up to at least one hundred degrees above the maximum temperature at which it is formed. This is regarded by some as another species, γ_2. Molecular pictures suggest that the γ-phase is molecular or a molecular H_2^+ ion held above the true chemisorbed species, the γ_2 being possibly held more strongly above vacant sites in the true chemisorbed monolayer. That it is a molecular species receives support from the observation that it does not undergo exchange reaction with deuterium. When hydrogen is adsorbed at temperatures above about 110°K, desorption rate measurements reveal two main discontinuities, one between 190 and 300°K and a second one in the region 300–600°K, the former designated the $\alpha_1\alpha_2$-phase and the latter the $\beta_1\beta_2$-phase. The heat of adsorption of the α-phase is $\sim$15 kcal mole^{-1} and is not as well defined as the two desorption peaks at the higher temperatures of 450°K and 500°K respectively termed the β_1- and β_2-phases. Of these, the β_2-phase is the most stable as it is the only phase formed if the adsorption takes place at temperatures above 473°K. It may be significant to note that the ratio of the two phases $\beta_1:\beta_2$ is found to approximate to 2:1 when adsorption takes place at 300°K, suggesting a crystallographic difference in structure. The bond strength of the strongly held hydrogen atom is $\sim$60 kcal mole^{-1}. The relation between the physically adsorbed, the γ- and the β_1,β_2-phases of chemisorbed hydrogen can be depicted as in Fig. 9. It is assumed that adiabatic dissociation of the hydrogen molecule results in two surface complexes formed by an electron passing into the vacancies in the d-band of the catalytic metallic substrate, the H_2^+ complex being held on empty atomic d-orbitals. For copper, gold and silver such vacancies as are necessary are attributed to d–s promotion. When we come to examine the catalytic behaviour of metals, of oxides and of doped oxides, the electronic configuration and the effectiveness of p-type semiconductors support the assumption that hydrogen is an electron donor. The change in surface potential on admission of hydrogen, however, indicates the passing formation of a complex that raises the work function, i.e. an effective dipolar array $Ni^+—H^-$; this decreases on continued adsorption and eventually a positive dipole $Ni^-—H^+$ or Ni^-HH^+ is formed. Consideration of the interatomic distances on the catalytic metals indicates that on face-centred nickel on the [110] planes the hydrogen atom can find a place in a plane below the top array of metal atoms. The ratio of atoms adsorbed on and in the

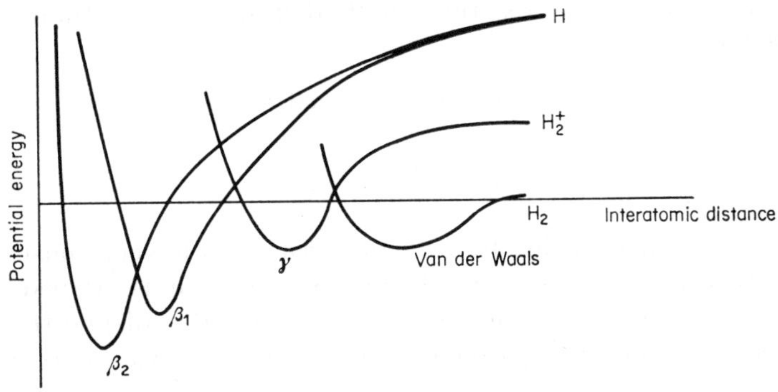

FIG. 9.

surface atomic spacings naturally varies from facet to facet but each is in a simple arithmetic ratio to another. The view that the atoms held in the lattice remove some conducting electrons to form hydride ions and those anchored on the lattice donate electrons to the substrate receives some support from the observed rise and subsequent fall in electric resistivity and in thermoelectric powers. We have already noted that there is still uncertainty as to the heat of adsorption of hydrogen at normal temperatures, but the data suggest that there is a change in the entropy of adsorption on progressive increase in coverage, an indication that at high coverages a certain amount of surface mobility is acquired, presumably in the β_2-phase, the γ-phase being readily mobile.

The Ortho–Para and Deuterium Exchange Reactions

Although the reaction kinetics of a catalytic reaction give us some information as to the processes involved and in certain instances it is possible to measure the individual energy terms comprising the apparent and true energies of activation of the surface reaction, the use of deuterium and other isotopes has permitted a great extension of our knowledge of the modus operandi of the reactions. As an example, it is found that ethane can be hydrogenated to methane at a nickel surface but, from the kinetics of the reaction alone, it is not possible to say whether the ethane is dissociated to two methyl radicals or to an ethyl radical and a hydrogen atom or, alternatively, whether the hydrogen is dissociated to atoms which in turn attack adsorbed ethane; also it is not possible without other information to define the rate-determining step in the reaction. Determination of the deuterium–hydrogen exchange reactions under various conditions (as well as the use of isotopic ^{14}C derivatives) provide us with additional information.

The basic exchange reactions with deuterium or tritium can be formulated

$$H_2 + D_2 \rightleftharpoons 2HD$$

$$H_2 + T_2 \rightleftharpoons 2HT$$

The information provided by exchange reactions is very considerably augmented if the ortho–para H_2 (or D_2) reaction is studied simultaneously as these latter reactions may be regarded as diagnostic for the presence of chemisorbed atomic hydrogen on catalysts that are not paramagnetic.

Paramagnetic action

The ortho–para hydrogen (or deuterium) reaction can be brought about by catalysts containing unpaired electrons. The paramagnetic conversion does not involve molecular dissociation into atoms and can thus be sharply distinguished from those catalytic systems that bring about the hydrogen–deuterium exchange reaction. At high temperatures, however, the hydrogen may undergo chemisorptive dissociation and the exchange mechanism may augment the ortho–para conversion. The paramagnetic conversion is considered to be due to a physically adsorbed molecule vibrating perpendicularly to the plane at the surface and at each vibration there is the possibility of a transition when the nuclear spin of the hydrogen molecule interacts with the inhomogeneous magnetic field of the catalyst surface: the probability of this event is given by

$$\phi = \frac{8\mu_a^2 \mu_p^2 I \pi^2}{9 r_s^6 h^2 kT} G(T)$$

where μ_a and μ_p are the magnetic moments of the paramagnetic site and proton respectively, I is the moment of inertia of the hydrogen molecule,

r_s is the distance of the hydrogen molecule from the magnetic centre and $G(T)$ is a function allowing for the equilibrium of the ortho–para mixture and the endothermicity of the transition. The absolute rate of conversion would then be

$$\omega = \phi n_0 \theta \nu$$

or, with a Langmuirian adsorption,

$$\omega = \phi n_0 \nu \frac{bp}{1+bp}$$

Where ν is the reciprocal of the mean life and n_0 is the number of paramagnetic sites per square centimetre. This spin isomerization reaction proceeds with no energy of activation and can be brought about by paramagnetic gases, e.g. oxygen or nitric oxide, by the ions of the rare earths and by metals which are paramagnetic and have surface free valencies due the the presence of unpaired electrons.

Atomic mechanisms in the ortho–para and exchange reactions

The chemical ortho–para conversion and the hydrogen–deuterium exchange reactions involve the participation of chemisorbed atoms. The original experiments of Bonhoffer and Farkas led to the conclusion that the mechanism involved adiabatic dissociation of a molecule on adsorption followed by interaction between two adsorbed atoms which evaporate in the molecular form. In this process partners can be exchanged. The reactions for the exchange reaction can be written

$$\mathrm{M+H_2 \rightarrow 2H(ads)}$$
$$\mathrm{M+D_2 \rightarrow 2D(ads) \qquad H(ads)+D(ads) \rightarrow HD}$$

and the ortho–para reaction can be written

$$\mathrm{M+o\text{-}H_2 \rightleftharpoons 2H(ads)+M \rightleftharpoons p\text{-}H_2+M}$$

or, alternatively, we may have a reaction involving two molecules of gas as in the exchange reaction.

An alternative mechanism in which a three-atom transition complex is formed by the interaction of a chemisorbed atom with a gaseous or physically adsorbed molecule has the advantage of economy in the breaking and the making of bonds.

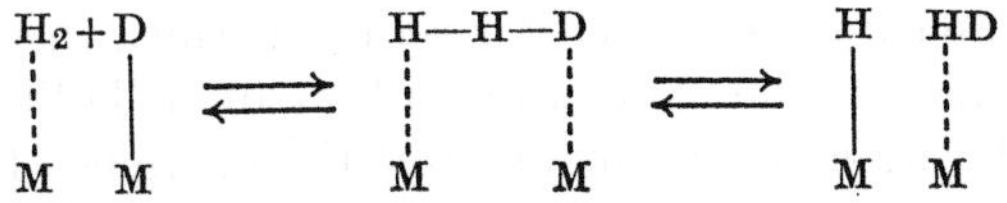

If it is assumed that the ortho–para conversion in the adsorbed layer is a first-order reaction with respect to molecular hydrogen, i.e. the recombination of atoms in the Bonhoffer-Farkas mechanism, or between a physically adsorbed molecule and a saturated atomic layer as in the Rideal mechanism, the first-order reaction constant is given by

$$k_1 = nkT(k_o + k_p)\frac{\theta}{p}\left(\frac{A}{V}\right)$$

where A/V is the ratio of catalyst surface to reaction volume and n is the number of sites per square centimetre, k_o and k_p being the monomolecular constants for the o $\rightarrow$ p and p $\rightarrow$ o reactions respectively. θ can be obtained from a simple Langmuir isotherm:

$$\theta = \frac{bp}{1+bp}$$

The relative rates of the ortho–para hydrogen and the ortho–para deuterium conversions and the exchange reactions are approximately:

$$\text{o–p}\, H_2 : \text{o–p}\, D_2 : H_2/D_2 :: 1 : 0{\cdot}54 : 0{\cdot}67$$

The zero-point energies have been evaluated and the minimum energy requirements for the reaction $D + H_2 \rightarrow DH + H$ is $0{\cdot}33 \pm 0{\cdot}02$ eV. The catalytic activity of the series of metals is found to be in the order

$$\text{Pt} = \text{Rh} = \text{Ru} = 10^2\text{Ni} = 10^2\text{Co} = 10^4\text{Fe} = 10^4\text{Cu}$$

and the energies of activation are approximately

$$E_{\text{Pt}} = 1, E_{\text{Ni}} = 3, E_{\text{Fe}} = E_{\text{Cu}} = 8, E_{\text{Au}} = 17{\cdot}5 \text{ kcal mole}^{-1}$$

The hydrogen–deuterium reaction also proceeds on a number of metallic oxides; of these, Cr_2O_3, Co_3O_4 and NiO are the most active. Less active are V_2O_3, MnO, Cu_2O and ZnO. Here it is found that doping with Ga^{3+} renders the oxide more active than doping with Li^+. The electronic factors round the metallic ion affect the catalytic activity more than the formal p- or n-type characteristics of the oxides. We thus have as active oxides with unpaired d-electrons: Cr_2O_3, $3d^3$; NiO, $3d^8$; Co_3O_4 (these are Co^{2+}, $3d^7$ and Co^{3+}, $3d^6$ ions). It appears that CoO has a very low catalytic activity, suggesting that the $3d^7$-ion is not active for the chemisorption of hydrogen. Other oxides with no or few unpaired d-electrons are not reactive. The vacancies in the d-band of a metal such as palladium can be filled by means of dissolved hydrogen, and in consequence the catalytic activity is reduced; thus, these reactions can be self-poisoning. In the same way, alloys of palladium and gold or copper and platinum can be made, and the energies of activation are found to rise as the electronic characteristics of the catalyst are changed.

Eley and Norton have made a careful examination of the three reactions on nickel over the temperature range 77–400°K. They find that the o–p H_2 and o–p D_2 reactions are closely similar over the range 77–300°K; the H_2/D_2 reaction is much slower but, on elevation of the temperature, it attains equality at 200°K. In the range 200–300°K the rates decrease in the order o–p $H_2 > H_2/D_2 >$ o–p D_2. The energy of activation for the H_2/D_2 reaction, $\sim 2{\cdot}4$ kcal mole^{-1}, rises rather rapidly to $\sim 7{\cdot}57$ kcal mole^{-1} at 270°K extending to 400°K. For the o–p H_2 reaction it is found to be 0 at 77–150°K, 1·54 kcal mole^{-1} at 200–300°K and $> 5{\cdot}0$ kcal mole^{-1} at 333–400°K.

That the two spin isomerisms proceed at low temperatures with almost identical rates and with zero energy of activation suggests that a paramagnetic mechanism is involved. At higher temperatures a chemical mechanism must be operative and, from the change in the H_2/D_2 reaction at about 270°K, clearly two different mechanisms must be involved. The velocity of the H_2/D_2 reaction is found to fall with decreasing pressure at this temperature, the $\log\omega$–$\log p$ plot giving values of 17·4 and 12·9 for $\log\omega$ at 1 and $\bar{3}$ for $\log p$ respectively. It may be concluded that this represents the operation of a Rideal mechanism, where the concentration of physically adsorbed molecules falls with decrease in pressure; the other reaction which is not so affected by pressure is believed to be a Bonhoffer-Farkas reaction taking place on the very active portions of the catalyst (some 1–2 % of the surface). Some support for the occurrence of both Rideal and Bonhoffer-Farkas reactions on the same catalyst is noted in experiments on exchange, for the complexities noted in the adsorption and desorption processes are reflected in the exchange reactions. If a deuterium covered surface is exposed to hydrogen, three stages in the rate of exchange are noted. Some of the gas exchanges very rapidly, some slowly and the remainder very slowly. It is claimed that these different rates cannot be due to heterogeneity of the surface, for different preparations of catalyst give the same or very similar results. It is possible that the very rapid reaction is due to the interaction of γ-adsorbed molecular hydrogen or molecular ions with chemisorbed deuterium in the α- and β-forms by a Rideal mechanism

$$H_2 + D(ads) \rightleftharpoons HD + H(ads)$$

The slower reaction can be ascribed to the action of a Bonhoffer-Farkas mechanism taking place with the β_1-deuterium.

$$\begin{aligned} H_2 &\rightarrow 2H(ads) \\ H(ads) + D(ads) &\rightarrow HD \end{aligned}$$

The very slow reaction may be due to the strongly held β_2-form involving atomic deuterium held below the surface level necessitating migration of chemisorbed hydrogen for interaction or displacement. We have noted that at not too elevated a temperature the o–p conversion proceeds at a greater rate than the deuterium exchange reaction; thus the dissociation and evaporation of deuterium cannot be the rate-determining step in the absence of a paramagnetic reaction. It is instructive to observe the effects of adding adsorbable deuterium-exchangeable poisons to the reacting system and to compare the relative rates of the o–p conversion and the deuterium exchange reaction. The data in Table XXXIV were obtained with a platinum catalyst at 17°C and

TABLE XXXIV

Poison added	Pressure (mm Hg)	Half-life Exchange	o–p conversion	Ratio
Ethyl ether	17	200	1	200
Benzene	40	13	0·5	26
Acetone	8	60	5	12
Water	12	13·5	7·4	1·8
n-Butanol	3	10	6	1·6
2-Ethylhexanol	1	10	4	2·5

20 mm Hg pressure of hydrogen or isotopic mixture. We note that the exchange reaction is much more sensitive to poisons than the o–p conversion, suggesting that the rate-determining step is that between the adsorbed poison radical and deuterium. Thus, with ammonia we might consider the following sequence of events:

$$NH_3 \rightarrow NH_2(ads) + H(ads)$$
$$NH_2(ads) + D_2 \rightarrow NH_2D + D(ads)$$

where the dissociative first step has an energy of activation of some 15 kcal mole^{-1} compared to 7·5 kcal mole^{-1} for the ortho–para conversion. The hydride dissociates into a hydrogen atom and a residue on the catalyst surface. When the o–p conversion is much faster than the exchange reaction the velocity of this exchange reaction is governed by the rate of dissociation of the hydride, but when the two reactions go at the same speed the rate of dissociation of deuterium is the rate-determining step. The strong adsorption of the hydride limits the extent of surface on which the deuterium molecule can dissociate, the exchange being controlled by the replacement of hydrogen atoms by deuterium.

The rate of adsorption to form a dissociated atom pair or radical and atom is given by the kinetic expression $\omega = Z\alpha(1-\theta)^2 e^{-E/kT} A$ where Z is the Hertz Knudsen collision number, α is the condensation coefficient and A is the area of the catalyst.

The Synthesis of Ammonia

Adsorption of nitrogen

As is the case with oxygen and hydrogen, various forms of adsorbed nitrogen have been noted which have been designated the α-, β- and γ-states. The α- and γ-phases, being low-temperature forms, are molecular since isotopic exchange does not take place with them. They differ in their nature: on tungsten, γ-nitrogen is desorbed below 200°K whereas the α-form is stable up to 400°K (in hydrogen adsorption the α-form refers to a high- and not low-temperature form). Also like hydrogen the β-form, which is stable up to higher temperatures, is composite in character, the two forms β_1 and β_2 being similar in that their stability temperature regions overlap one another, and both permit isotopic exchange. The condensation coefficient of nitrogen is also low, being $\sim 0{\cdot}3$ on polycrystalline tungsten. The β_1-phase desorbs with first-order kinetics but the β_2-form evaporates according to a higher order.

In field-emission experiments, nitrogen is not adsorbed on the [110] planes of tungsten at 300°K, although the other facets are densely populated with atoms on exposure to gaseous nitrogen. The temperature has to be brought down to 190°K for condensation to occur on this plane. On the other planes, the heat of adsorption is strongly exothermic, $\Delta H = -3{\cdot}7$ eV and is practically irreversible. Ehrlich, to whom most information is due, calculated that at 300°K and at a coverage of 7×10^{12} atoms cm^{-2} the equilibrium pressure would be 10^{-8} torr or less if the energy of bonding W—N exceeded $\Delta H = 5{\cdot}3$ eV. At 145°K this coverage could be reached with a value for $\Delta H = 5{\cdot}1$ eV. This falls below the value for the other facets where $\Delta H = 6{\cdot}7$ eV.

Nitrogen can be readily chemisorbed in the molecular α- and γ-forms on metals having two vacancies in the d-band or dsp-hybrid orbitals, e.g. Pt, Rh, Pd. Adsorption on nickel is weak and molecular. Adsorption can also take place on oxides such as Cr_2O_3. Various structures have been proposed for these chemisorbed molecular forms, one a π-bonding

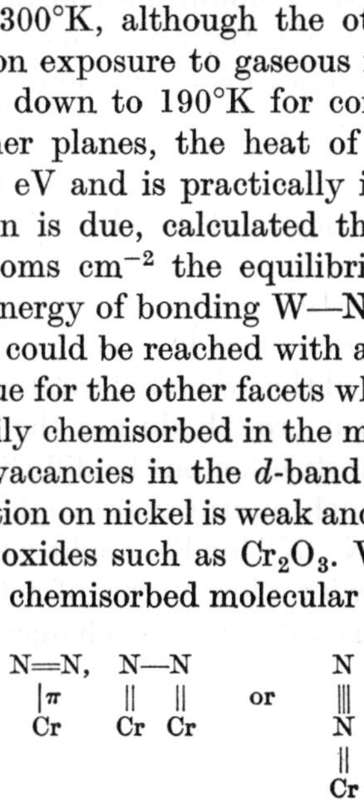

The last form has been suggested since some slight differences in the infrared absorption spectra of $^{15}N{\equiv}^{14}N{=}$ and $^{14}N{\equiv}^{15}N{=}$ have been claimed.

Chemisorption of nitrogen in the β- or atomic forms takes place most readily on metals that have three or more vacancies in the d-band, such as W, Ta, Mo, Ti, Zr, Fe. We may also include elements such as Ca and Ba. Iron will adsorb nitrogen at low temperatures with a heat evolution of 10 kcal mole^{-1} at $\theta = 0$ to 5 kcal mole^{-1} at $\theta = 1$. There is also an activated adsorption of nitrogen where the heats of adsorption vary from 70 to 40 kcal mole^{-1} at $\theta = 0$ to 16 kcal mole^{-1} at $\theta = 1$. It has been claimed that at $\theta = 1$ one nitrogen atom requires five iron atoms. This, however, may be transitory since nitrogen can dissolve in α-iron up to a quantity represented by Fe_2N. On heating such nitrided iron molecular nitrogen is liberated according to a bimolecular reaction. The apparent fall in ΔH with coverage may be due to the effects of bulk solution rather than surface interaction. Isotopic nitrogen exchange can readily take place at 250°C on osmium, and at about 450°C on molybdenum and on promoted iron catalysts, but over 1100°K is required to bring about exchange on tungsten. Promoted iron catalysts used in the ammonia synthesis are generally prepared by reduction in hydrogen at 500°C of a mixture 95 % Fe_3O_4 plus 4–5 % Al_2O_3 plus 0–1 % K_2O.

The necessity for using catalysts active at low temperatures is evident from the magnitude of the equilibrium constant of the reaction; very small values obtaining at temperatures above 400°C. The equilibrium constant is given by the following expression

$$\tfrac{1}{2}N_2 + \tfrac{3}{2}H_2 = NH_3(g)$$

$$\log_{10} K = \frac{2075}{T} - 2{\cdot}50 \log T - 1{\cdot}26 \times 10^{-4} T + 0{\cdot}19 \times 10^{-6} T^2 + 2{\cdot}10$$

Thus at 298°K, $K = 3{\cdot}4 \times 10^2$.

The function of the promoters has not been completely elucidated. One important property of the promoted catalyst is the restriction in the rate of crystal and grain growth during the reaction at 450–500°C, presumably due to the restriction on the diffusion of iron atoms caused by spinel barriers. All such mixtures are copious positive ion emitters and ion exchange of $-N{\equiv}N^+$ for K^+ might well provide a mechanism for nitrogen "fixation" with a low energy of activation. Three different versions have been proposed for the exchange mechanism and can be depicted as follows.

$$^{14}N(ads) + {}^{15}N_2(g) \rightleftharpoons {}^{14}N^{15}N_2(ads) \rightleftharpoons {}^{15}N(ads) + {}^{14}N^{15}N(g) \quad \text{(a)}$$

$$^{14}N(ads) + {}^{15}N(ads) \rightleftharpoons {}^{14}N^{15}N(g) \quad \text{(b)}$$

corresponding to Rideal and Bonhoffer-Farkas mechanisms respectively, and in addition a doublet mechanism has been proposed

$$^{14}N_2 + ^{15}N_2 \rightleftharpoons \begin{array}{cc} ^{14}N\text{—} & ^{15}N \\ | & | \\ ^{14}N & ^{15}N(\text{ads}) \end{array} \rightarrow 2\,^{14}N^{15}N \qquad \text{(c)}$$

Since the β_1-phase desorbs with first order kinetics the desorbed molecules must arise from a surface complex such as exists in (a) or (c) but not in (b). The β_2-phase decomposition, which proceeds with kinetics of a higher order and more slowly, may well indicate a Bonhoffer-Farkas mechanism (b) and also indicate that the surface atoms in this state are mobile. It has been said that gaseous nitrogen can exchange with chemisorbed nitrogen at temperatures as low as 300°K.

Adsorption of ammonia

Ammonia is readily physically adsorbed. Rhodin examined adsorption on crystals of copper and found that at $\theta = 0$ the heats of adsorption on the [110], [100] and [111] planes were practically identical $\sim 2{\cdot}0$ kcal mole^{-1}, where dipolar repulsion prevented the physical attractive forces from effecting two-dimensional condensation which sets in at higher pressures and coverages with $\Delta H \sim 2{\cdot}7$ kcal mole^{-1}. He found that on the three planes the heat of adsorption was highest on the plane most closely packed. The gas is chemisorbed on the catalytically active metals with decomposition. The amount adsorbed at $\theta = 1$ is equal to the amount of hydrogen that can be chemisorbed, and thus ammonia occupies two sites per molecule. The primary action of iron is dissociative, and some hydrogen appears in the gas stream on desorption.

$$2M + NH_3 \rightarrow MNH_2 + MH$$

On the addition of excess ammonia, hydrogen is displaced

$$MH + NH_3 \rightarrow MNH_2 + H_2$$

The complex MNH_2 breaks down slowly, and eventually iron nitride is formed. We have already noted a rough parallelism between the energies of activation for decomposition and the work functions of the various catalytic metals, suggesting that nitride decomposition is the rate-determining step. On many metals the reaction is of zero order with respect to ammonia, but reverts to first order on platinum. Hydrogen may be an inhibitor but since it is found that NH_3 decomposes more rapidly than ND_3 the zero order in the reaction must be due to an inhibition, e.g. $\omega \propto p_{NH_3}/(1+\theta_{\text{nitride}})$, where the evaporation of nitrogen is the rate-determining step.

The experimental results suggest that the nitride formation may not be confined to the surface but that the total amount on and just below the surface is determined by the necessity for the evaporation rate to become equal to the rate of formation. Experimental work on tungsten has confirmed that at the same nitride concentration the rate of its formation from ammonia is equal to the rate of desorption of nitrogen.

Synthesis of ammonia

Although the rate of production of ammonia from synthesis gas is comparable with that of nitride formation and exchange reactions involving deuterium, hydrogen or deutero-ammonia and ammonia take place at far lower temperatures, the rate of nitride formation is not necessarily the limiting factor in ammonia synthesis, for it is found that if deuterium be substituted for hydrogen the production of deutero-ammonia is faster than that of hydrogen ammonia, suggesting that the rate is governed by the reaction between a chemisorbed nitrogen or imide and gaseous hydrogen. Light is thrown on the mechanism by consideration of Horiuti's concept of the stoichiometric number.

The stoichiometric number is the number of times the slowest step is involved in the complete overall reaction. We have seen from decomposition experiments that a number of steps are involved in the decomposition of ammonia, and we may anticipate that these will constitute the mechanism of the reaction $N_2 + 3H_2 \rightarrow 2NH_3$. The series of steps is consequently:

$$\begin{aligned} N_2 &\rightleftharpoons 2N(ads) & n &= 1 \\ H_2 &\rightleftharpoons 2H(ads) & n &= 3 \\ N(ads) + H(ads) &\rightleftharpoons NH(ads) & n &= 2 \\ NH(ads) + H(ads) &\rightleftharpoons NH_2(ads) & n &= 2 \\ NH_2(ads) + H(ads) &\rightleftharpoons NH_3 & n &= 2 \end{aligned}$$

where n is the stoichiometric number. Clearly, if the chemisorption of nitrogen is the slow step the stoichiometric number is one, and it will be two if either amide or imide formation is rate-determining. By studying the exchange reaction between nitrogen isotopes and nitrogen with ammonia isotopes in a flow system where the gases have attained equilibrium, together with separate investigations on the rates of chemisorption and desorption of nitrogen in the presence and absence of hydrogen, the individual steps can be isolated. If in addition we examine the rate of ammonia formation from hydrogen and chemisorbed nitrogen as a function of the partial pressures of the gases and the coverage, it is possible to evaluate n.

It has been concluded that at low temperatures and at high pressures of hydrogen the rate of chemisorption of nitrogen is rate determining ($n=1$) but at higher temperatures the rate of hydrogenation of the nitride governs the reaction ($n=2$). We can depict these two mechanisms in the following way. At low temperatures and at high hydrogen pressures the rate of reaction is governed by the rate of nitride formation, and the subsequent reactions are presumed to occur rapidly and can be considered to be readily reversible; hence

$$N_2 \longrightarrow N(ads) \overset{H_2}{\rightleftharpoons} NH(ads) \overset{H_2}{\rightleftharpoons} NH_3(ads) \rightleftharpoons NH_3$$

The rate of nitride formation (and hence the rate of reaction) is

$$\omega_N = Kp_{N_2}(1-\theta_N)^2 \tag{10}$$

where θ_N is the fraction of the surface covered with nitride. Since the subsequent reactions are in reversible equilibrium we can write

$$\theta_N + H_2 \rightleftharpoons NH_3$$

If the equilibrium constant of this reaction is K we have

$$\theta_N = \frac{Kp_{NH_3}}{p_{H_2}^{3/2} + Kp_{NH_3}}$$

and we find for the rate of reaction (10)

$$\omega = \frac{K'' p_{N_2} p_{H_2}^3}{(Kp_{NH_3} + p_{H_2}^{3/2})^2}$$

At lower pressures of hydrogen, since the reaction proceeds more rapidly with deuterium than with hydrogen, the slow step is in all probability the first introduction of hydrogen into the nitride; the subsequent further addition of hydrogen to form ammonia proceeding readily with an energy of activation E of only some 3·5 kcal mole^{-1}. The scheme of reaction on this view would be

$$N_2 \rightleftharpoons N(ads) \xrightarrow{H_2} NH(ads) \overset{H_2}{\rightleftharpoons} NH_3(ads) \rightleftharpoons NH_3$$

The rate of reaction is accordingly

$$2\omega = kp_{H_2}\theta_N^2$$

The surface nitride concentration is regarded as in dissociative equilibrium with the nitrogen gas and, consequently,

$$\nu\theta_N^2 = kp_{N_2}(1-\theta_N-\theta_{NH})^2$$

The surface imide can also be regarded as in equilibrium with the ammonia or

$$\theta_{NH} + H_2 = K' p_{NH_3}$$

or

$$\theta_{NH} = \frac{K' p_{NH_3}}{p_{H_2} + K' p_{NH_3}}$$

We thus obtain for the rate of ammonia formation

$$2\omega = K' p_{H_2} p_{N_2} \left\{ \frac{p_{H_2} + K' p_{NH_3}}{(1+\theta_N)(p_{H_2} + K' p_{NH_3}) + K' p_{NH_3}} \right\}^2$$

This somewhat complex expression reduces at low temperatures when the rate of imide formation is small, i.e. $\theta_N \sim 1$, to

$$2\omega = K' p_{H_2} p_{N_2} \left\{ \frac{p_{H_2} + K' p_{NH_3}}{2p_{H_2} + 3K' p_{NH_3}} \right\}^2$$

In adopting the hypothesis of equilibrium of the ammonia gas with the adsorbed reactant there must be a concomitant decomposition of the gaseous ammonia, which we have seen can be written

$$-\bar{\omega}_{NH_3} = \frac{K p_{NH_3}}{1+\theta_N} = K \frac{p_{NH_3}^2 + K p_{NH_3} p_{H_2}^{3/2}}{p_{H_2}^{3/2} + 2K p_{NH_3}}$$

so the net rate of production of ammonia will be $\omega - \bar{\omega}_{NH_3}$.

Tamaru has applied the principle of detailed balancing to the synthesis of ammonia from its elements. In the steady state the net rate of ammonia formation from hydrogen and chemisorbed nitrogen plus the rate of desorption of nitrogen must be equal to the rate of chemisorption of the nitrogen plus the rate of nitride formation resulting from the decomposition of the ammonia. He considers that all the reactions follow the kinetics of activated processes with energies given by the Elovitch equation. He found that both the adsorption of nitrogen and its desorption in the presence of hydrogen did indeed so conform. He writes for these reactions

$$\overrightarrow{\omega_1} = A p_{N_2} e^{-\alpha\theta_N}$$

$$\overleftarrow{\omega_1} = B e^{\beta\theta_N}$$

where A, B, α and β are constants. The rate of reaction between the surface nitride and hydrogen to yield ammonia is assumed to take the form

$$\overrightarrow{\omega_2} = C p_{H_2} e^{\gamma\theta_N}$$

whereas the decomposition of the gas at the metal surface following the kinetics obtained with tungsten is

$$\overleftarrow{\omega}_2 = \frac{D p_{NH_3} e^{-\delta\theta_N}}{p_{H_2}^{1/2}}$$

The rate of the net production of ammonia is evidently the difference between the quantity of ammonia formed and that lost or between the quantity of nitrogen adsorbed and desorbed in the same time, or

$$\omega - \bar{\omega}_{NH_3} = \overrightarrow{v}_1 - \overleftarrow{v}_2 = \overrightarrow{v}_2 - \overleftarrow{v}_1$$

Further if, as has already been assumed, the ammonia is in equilibrium with the reactants we have

$$1 = \frac{\omega_1}{\overrightarrow{\omega}_2} = \frac{\overrightarrow{v}_2}{K_p}$$

where

$$K_p = p_{NH_3}^2 / p_{H_2}^3 p_{N_2}$$

The two constants in the Elovitch equation are related to one another as follows.

$$2(\gamma+\delta) = \alpha+\beta \quad \text{and} \quad \frac{A}{B} = \left(\frac{D^2}{C^2}\right) K_p$$

Tamaru obtained the following values for these constants on a promoted iron catalyst: $\alpha = 0{\cdot}177$ ml^{-1}; $\beta = 0{\cdot}106$ ml^{-1}; $\gamma = 0{\cdot}096$ ml^{-1}; $\delta = 0{\cdot}046$ ml^{-1}; $A = 20{\cdot}4$ ml N_2 min^{-1}; $B = 5{\cdot}0 \times 10^{-4}$ ml N_2 min^{-1}; $C = 0{\cdot}91 \times 10^{-4}$ ml $NH_3/2$ cm^{-1}; $D = 8{\cdot}1$ (ml $NH_3/2$) cm^{-1}. Since it would appear that $\beta \sim \gamma$ we find

$$\frac{\overleftarrow{\omega}_1}{\overrightarrow{\omega}_2} = \frac{K e^{(E_2 - E_1)/RT}}{p_{H_2}}$$

where E_2 and E_1 are the activation energies for $\overrightarrow{\omega}_2$ and $\overleftarrow{\omega}_1$ respectively; the latter is equal to the sum of the heat of adsorption of nitrogen and its energy of activation for desorption and is much larger than the energy of activation for the reaction between hydrogen and nitride. At a pressure of hydrogen where $\overleftarrow{\omega}_1/\overrightarrow{\omega}_2 = 1$ (between 1 and 10 cm Hg) the reaction path is a consecutive one, with two free energy steps.

$$N_2 \rightarrow N(\text{ads}) \rightarrow NH$$

Catalysis at Electrodes

We have already noted that heterogeneous catalytic reactions involve electronic transfer and, since electrodes serve as electron sources and sinks, we might anticipate that, in addition to the control of the electron affinity of the acceptor by the electrode potential, the nature of the electrode would play some part in a reaction taking place at its surface. In some cases both the nature of the reaction as well as the efficiency of formation of the product are dependent on the choice of the electrode. Evidently, a limit is set to the permissible current density by the supply of electron acceptors or electron donors at the interface. Limitations imposed by diffusion and the effects of movement of the electrolyte relative to the electrode on the thickness of the diffusive layer have been dealt with by Nernst, Sand and Levitch.

We shall have occasion to note that the catalytic action of the electrode has been confirmed not only in the efficiency of the electrolytic production of hydrogen and oxygen but also in processes of electrolytic reduction and oxidation. Thus Haber showed that by suitable choice of cathode, nitrobenzene would be reduced in turn to azoxybenzene, azobenzene, hydrazobenzene and aniline.

On placing a metal in a solution containing its ions a particular ionic arrangement is set up in the neighbourhood of the electrode and the extent of adsorption of any non-electrolyte, if present, is dependent on the ionic composition of the surface phase. There will also come into existence a potential difference between the electrode and the solution. We can express the conditions of equilibrium in terms of the electrochemical potential μ_{M^+} of the metallic ions in the two phases, the electrode and electrolyte, and find

$$FnV + k_m = RT \log a + k_u$$

Where F is the Faraday constant, n is the valency of the ion and V is the potential difference between the electrode and the electrolyte of ionic activity a. Here, k_m and k_u are factors representing the effects of environment of the ion in the metal and solution respectively. Nernst regarded the process by which the potential difference between a silver electrode and electrolyte arose as due to silver ions going into solution to form hydrated ions, leaving electrons in the metal. Thus the process of ionic solution continues until a double layer of electrons in the silver and a double layer of positive ions in the solution are built up of such a magnitude that equilibrium is established between the rates of solution of silver as ions and of the redeposition of silver ions as metal. This latter is caused by the electrostatic attraction of the negative charge of the double layer. This is Nernst's ionic transfer process sometimes expressed as the electrolytic solution pressure of the metal. There is thus a critical potential difference between a metal and a characteristic concentration of the ions. Evidently $k_m - k_u$ can be represented as $RT \log a_0$ the logarithm of Nernst's ionic equilibrium. Frumkin was the first to attempt an evaluation of a_0 by determining the solubility of silver ions when silver foil is brought into contact with water.

Attempts have been made to arrive at the free energy change in the reaction from the individual steps involved, namely the latent heat of evaporation and the work function of the metal, the ionization potential and heat of solvation of the ion together with the change in entropy involved in the reaction. In general, since single electrode–electrolyte potential differences are difficult to measure or calculate, use is made of standard half elements to which reference can be made. The hydrogen electrode is taken as the arbitrary zero in the electromotive scale, the calculated value of which is of the order of 0·46 volt.

At equilibrium the rate of discharge of ions into the metal is equal to the rate at which metal atoms undergo ionization and solution, i.e. there is an exchange current such that $\overrightarrow{\omega} = \overleftarrow{\omega}$. In order to cause a net discharge of ions the potential of the electrode with respect to the solution has to be raised above the reversible potential, this excess is the overpotential. Matsuda has examined the case of hydrogen discharge at a metal surface where the overpotentials η for different current densities have been examined in detail. The overpotential can be defined as the difference in chemical potential of the metal electron $\Delta\mu_e$ between the reversible and irreversible states of the hydrogen electrode or

$$\eta = -\frac{\Delta\mu_e}{F}$$

The chemical potential of the metal electron is made up of an electrostatic potential χ and the electronic work function of the metal ϕ.

$$\mu_e = -F(\chi+\phi)$$

whence

$$\eta = \Delta\chi+\Delta\phi.$$

These two contributions to the overpotential can be depicted as in Fig. 10.

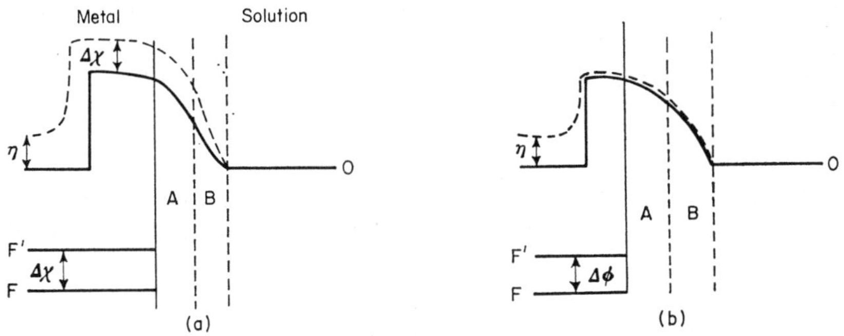

FIG. 10.

The Fermi level F of the metal electrons is raised as the surface becomes negatively charged. Any change in contact potential between the metal and its outer metallic connection causes a change $\Delta\phi$ in work function, a contributor to the overpotential; likewise, the electric double layer at the interface, where A is the Helmholtz layer and B the Gouy layer, increases as the charge increases, causing a change in the metal solution potential difference $\Delta\chi$ and this also contributes to η.

A great deal of information concerning hydrogen discharge mechanisms can be obtained from overvoltage–time curves in which measurements are made of the rate of rise of overvoltage and its rate of decay as a function of the amount of current passed in short intervals of time.

If a small quantity of current i is passed, the double layer charge is increased and later intermediate products are built up. In the initial stages, however, the double layer charge is increased but at the same time it loses charge by reason of Faradaic leakage due to the existence of the exchange current. Consequently, if a quantity of electricity $q = i\delta t$ is passed, the surface charge will increase by an amount q' but suffer a loss q'' by reason of the exchange current. We have the following relationships

$$q = q' + q''$$

$$q = i\delta t$$

$$q' = C\,\mathrm{d}V$$

$$q'' = i_0\,\delta t$$

where C is the capacity of the double layer and i_0 the exchange current. Hence, we obtain for the exchange current

$$i = i_0 + C\frac{\partial V}{\partial t} = i_0 + C\frac{\partial \eta}{\partial t}$$

the small changes in η being accounted for by changes in the surface potential χ with no significant change in the work function of the metal. If the electrode is charged up and the magnitude of the exchange current, i.e. the rate of decay of the overvoltage measured, we have

$$C\left(\frac{\partial \eta}{\partial t}\right)_{t=0} = -i_0$$

a relationship giving us a convenient method, first used by Frumkin, for measuring the capacity. The capacity of the double layer is made up of the two capacities, the Helmholtz and Gouy layers respectively, i.e.

$$\frac{1}{C} = \frac{1}{C_{\text{Helmholtz}}} + \frac{1}{C_{\text{Gouy}}}$$

Since the capacity of the Gouy layer is relatively large, the term $1/C_{\text{Gouy}}$ can be neglected and the changes ascribed to the Helmholtz layer alone.

By this procedure it is possible to determine the magnitude of the exchange currents not only for gases but also for metals in solutions of their ions at their reversible potentials. Evidently, specific surfaces of metals can also be evaluated provided that the data for a uniform smooth surface are available for comparison and that there is no sensible change in the capacity of the double-layer. Bowden and Rideal used this method to determine the specific surfaces of a number of catalytic metals. Table XXXV lists the data they found for nickel.

TABLE XXXV

Treatment		Specific surfaces
Activated	New	46
	Old	29
Annealed	New	10·5
	Old	7·7
Electroplated	New	12
	Old	9·5
Rolled	New	5·8
	Old	3·5
Polished	New	13·3
	Old	9·7

Although in the case of hydrogen at metal surfaces in both acid and alkaline solutions at constant ionic strength the capacity $C(=18\,\mu\text{F cm}^{-2})$ does not change over a small range of voltage, this is not the case with electrolytes in general. The variation of capacity with electrolyte and surface charge has been investigated in detail using the above method by Frumkin, Grahame and others. Thus, with sodium fluoride at a concentration of 10^{-5} mole cm^{-3} values of $C = 23\ \mu\text{F cm}^{-2}$ with a surface charge of $q' = 8\ \mu$ coulombs cm^{-2} falling to a minimum of $11\ \mu\text{F cm}^{-2}$ at $q' = 0$ and rising to some $16\ \mu\text{F cm}^{-2}$ at a value of $q' = 4\ \mu$ coulombs cm^{-2} were obtained.

The exchange current or the velocity of the exchange reaction at the potential of the reversible electrode, e.g. the rate which silver ions present in the Helmholtz double layer are discharged onto a silver electrode, or the reverse process, can be regarded as a unimolecular reaction, i.e.

$$v = i_0 = \kappa e C\delta \frac{kT}{h} \mathrm{e}^{-\Delta G\ddagger/RT}$$

where κ is the transmission coefficient (usually 1), $C\delta$ represents the ionic concentration in the Helmholtz layer (C being the bulk concentration and δ the double layer thickness), e is the electronic charge, $\Delta G^{\ddagger}$ the free energy change in forming the activated transition complex between the ion and the metal. If we depart from the reversible potential and a potential difference $\Delta\chi$ is formed at the interface, the concentration of the ions in the Helmholtz layer will alter and attain a value $C\delta\,\mathrm{e}^{-\Delta\chi e/RT}$. The velocity of reaction will accordingly become

$$\vec{\omega} = i = \kappa e C\delta \frac{kT}{h} \exp\left(-\frac{\Delta G^{\ddagger}}{RT} - \frac{\Delta\chi e}{RT}\right) = hC\,\mathrm{e}^{-(\Delta G + \Delta\chi e)/RT}$$

In this expression we have assumed that the surface concentration of silver ions (θ_{Ag^+}), i.e. those in the Helmholtz layer, is directly proportional to the bulk concentration. We should anticipate on analogy with, say, potassium on tungsten that this would not be the case and that the relative amounts of chemisorbed ions and of water would vary with the coverage. For example, Frumkin has shown that the relative amounts of chemisorbed hydrogen and water vary with $\Delta\chi$ such that

$$\frac{-\mathrm{d}\theta}{\mathrm{d}\Delta\chi} = \frac{2e}{r}$$

where r is given by $\Delta H_{\theta} = \Delta H_{\theta=0} - r\theta$.

The exchange currents for different electrode systems have been determined by numerous investigators, notably by Frumkin, Tanada, O. M. Bockris and Spiro. The standard conditions adopted are usually 25°C and a molar electrolyte; the exchange current is expressed in amperes cm^{-2} at the reversible electrode potential. Some values are given in Table XXXVI.

TABLE XXXVI

System	i_0 (A cm^{-2})	System	i_0 (A cm^{-2})
Ag in NH_4ClO_4 with 0·1 M Ag^+	2·4 to 5	Cu in 0·1 N HCl	2×10^{-7}
		Cu in 0·15 N NaOH	1×10^{-6}
Cd in 0·8 N K_2SO_4 with 10^{-2} M Cd^{2+}	$1{\cdot}5 \times 10^{-3}$		
Hg in 0·98 M $HClO_4$ with mM Hg_2^{2+}	0·25		
Ni in M $NiSO_4$	$10^{-8{\cdot}7}$	Ni in 0·12 N NaOH	4×10^{-7}
Ni in 0·5 M $NiSO_4$, p_H 7·9	10^{-5}		
Ni in 0·5 M $NiSO_4$, p_H 5·5	10^{-7}	Zn in M $ZnSO_4$	$10^{-4{\cdot}7}$

Information on electrode mechanisms, e.g. the reduction of a non-electrolyte at a metallic cathode can also be obtained from consideration of the behaviour of the mercury capillary electrode. If the mercury is in contact with an acid solution of its sulphate, the individual independent variable components are Hg^{2+}, SO_4^{2-}, H^+, OH^- and electrons. If there is a change in γ the free surface energy of the metal electrolyte interface and the strength of the sulphuric acid be kept constant there will be no change in the concentration of SO_4^{2-}, H^+, or OH^-. Utilizing Gibbs' relationship for the change in free surface energy

$$\delta\gamma = -\sum \Gamma \,\mathrm{d}\mu$$

Where Γ is the surface excess of a component, we can choose the dividing surface in a manner to maintain Hg^{2+} constant and thus have only the electrons to consider or

$$\Gamma_{\text{electrons}} = \left(\frac{\mathrm{d}\gamma}{\mathrm{d}V}\right) = -q$$

We can, in this manner, derive Lippmann's equation and determine the surface charge and the number of gram equivalents of mercurous ions taken on by an expanding mercury surface or thrown off by a contracting one in the course of Nernst's ionic transfer.

More recently, Frumkin, Petric and Marvet have examined the behaviour of a hydrogen electrode in a similar manner. Although it is impossible to measure directly the free surface energy of the platinum and its charge as a function of Γ_{H^+} it is possible to derive expressions for $\mathrm{d}V/\mathrm{d}\mu_{H^+}$ with $\Gamma_{H^+} = q$ the free surface charge and Γ_H the hydrogen adsorbed in the atomic form.

The differential electrical capacity C of the double layer is evidently

$$\frac{-\mathrm{d}q}{\mathrm{d}V} \text{ or } C = \frac{\mathrm{d}^2\gamma}{\mathrm{d}V^2}$$

Thus with a constant capacity the γ, V curve should be parabolic in form having a superficial charge density of zero at γ_{max}. Departures from the parabolic form indicate that specific adsorption of some constituents at the interface is taking place. If ionic, the specific adsorption is enhanced for anions when the applied potential renders the mercury positive and vice versa.

Neutral molecules, e.g. n-heptanol are appreciably adsorbed near the maximum of the capillary curve but are displaced by preferential adsorption of ions on either side of the maximum. Aromatic compounds are adsorbed on the positive side of the capillary curve since their π-electrons interact with the positively charged surface, and the adsorption of polarizable molecules can also be altered by departures on the positive or negative sides of the curve. Such adsorption is reflected in changes in capacity of the double layer. Thus, on the addition of n-heptanol, the capacity of the double layer is reduced to some 3·6 μF cm^{-2}. This may be taken to represent a complete monolayer of heptanol, and a rough estimate of the quantity of heptanol adsorbed from a dilute solution can be evaluated from the change in capacity of the double layer. We are indebted to Grahame and especially to Frumkin for unravelling the somewhat complex factors that affect the relative adsorptions as the electrode potential relative to the solution is changed.

The intrusion of catalytic action by the electrode material was first noted by Tafel in the discharge of hydrogen ions. Tafel found that the overpotential necessary to bring about an increase in current and rate of gas liberation was characteristic of the metal and given by the following relationship

$$i = i_0 e^{\alpha F \eta / kT}$$

where α is the Tafel constant and i_0 the exchange current at the potential of the reversible hydrogen electrode. Not only have metals characteristic values for i_0 but they can also be grouped into three sets which have values for the Tafel constant α of 0·5, 1·5 and 2 respectively. High overpotentials were found for metals such as Hg or Pb and low overpotentials for the efficient hydrogenating catalysts such as Ni or Pt. In some cases, the Tafel constant changes from one value to another at a critical current density.

There are evidently several mechanisms by which hydrogen in the form of the solvated proton H_3O^+ or HOH can appear eventually as hydrogen gas after the passage of an electron across the metal–liquid interface. We are indebted to Frumkin and especially to Horiuti for identifying the separate steps in the complex series constituting the discharge mechanism and for the assignment of the slow or rate-controlling steps in the different cases. The three mechanisms have been termed the slow discharge, the catalytic and the electrochemical respectively.

In the steady state of gas evolution the surface at the electrode contains H^+, H, H_2 and probably H_2^+ as constituents, where H^+ is associated with the Brønsted base H_2O as H_3^+O or with OH^-. The various possible reactions taking place are

$$H^+(ads) + e \rightarrow H(ads) \quad (11)$$

$$2H(ads) \rightarrow H_{2a} \rightleftharpoons H_2(g) \quad (12)$$

$$H^+ + H(ads) + e \rightarrow H_2(ads) \rightleftharpoons H_2(g) \quad (13)$$

$$2H^+ + e \rightarrow H_2^+(ads) \quad (14)$$

$$H_2^+(ads) + e \rightarrow H_2(ads) \rightleftharpoons H_2(g) \quad (15)$$

The slow discharge mechanism has (11) or the passage of the electron from the metal as its rate-controlling reaction; in reaction (12), viz. $2H(ads) \rightleftharpoons H_2(ads) \rightleftharpoons H_2(g)$, all three reactants are in reversible equilibrium. In the catalytic mechanism the rate-controlling step is (12) and, consequently, the atomic hydrogen on the surface, H(ads), increases with increase of current, whereas the electrochemical mechanism involves either one or all of the steps under (13), (14) and (15) as rate controlling. Futher variants of (13), the electrochemical or ion–atom reaction step, might be suggested in the forms

$$H^+(ads) + H(ads) \rightarrow H_2^+(ads) \quad (16)$$

followed by (15) or

$$H^+ + H(ads) \rightarrow H_2^+(ads) \quad (17)$$

since if H_2^+ exists as an intermediary it is evidently more readily formed by a Bonhoffer–Farkas (16) or Rideal (17) mechanism than by the termolecular reaction (14). Similar considerations are applicable to (13). The slow discharge mechanism results in a value of $\alpha = 0{\cdot}5$, for the catalytic mechanism $\alpha = 2$ and for the electrochemical $\alpha = 1{\cdot}5$. The value of $\alpha = 2$ for the catalytic mechanism rests upon the premise that the rate of desorption of hydrogen from the surface-adsorbed hydrogen atoms proceeds by a bimolecular process or $i = K\theta_H^2$, where θ_H is the fraction of the surface covered. If there are repulsive forces between the adatoms such that the free energy F took the form $F = RT \log \theta$, Horiuti showed that the Tafel constant would be 0·5 instead of 2.

TABLE XXXVII

η (mvolts)	0	50	120	180	240	480	1000
θ	0·31	0·36	0·44	0·46	0·51	0·80	0·97

Computation of θ for various values of η on nickel at 25°C in slightly acid solutions are given in Table XXXVII. Bockris has considered in detail the factors affecting the catalytic activity of the electrodes in hydrogen evolution. In general, we note that if the gas is strongly adsorbed, i.e. $D_{M—H}$ has a high value, the proton discharge reaction is fast, i_0 is consequently large and the combination reaction consequently is rate controlling. On mercury, on the other hand, $D_{M—H}$ is small and the slow discharge mechanism is rate controlling. Indeed there is a linear decrease in η (at 10^{-3} amp cm^{-2}) with increasing $D_{M—H}$ for Pb, Hg and Ti and a linear increase in η with increasing $D_{M—H}$ for many other metals. There is always uncertainty as to the relationship between the bond strength (D_{M-H}) and η–D. Gold, silver and copper are found to give values close to Pt, Ni and Fe. The dependence of i_0 on the strength of the M—H bond is also revealed from a study of palladium–gold alloys. On progressive filling up of the d-band, with gold the M–H bond strength and i_0 decrease until the completion at 60% limit. Again, when the metal work function ϕ is plotted against $\log i_0$, the exchange current for hydrogen, a similar differentiation between the two groups of metals is found.

In electrochemical reduction of organic substances, e.g. nitrobenzene (N), it is clear that reaction takes place most efficiently when, of the two competing mechanisms $N + H(ads) \rightarrow NH$ and $2H(ads) \rightarrow H_2$, the former reaction is favoured; thus, strong adsorption and electrodes in which the slow discharge mechanism is operative or electrodes with small values of i_0 capable of fairly high overpotentials will be most effective.

Little is known of catalysis in electrochemical oxidation, since the electrode mechanisms are not known with any degree of certainty. Although the reversible electrode system

$$O_2 + 2H_2O + 4e \rightleftharpoons 4OH^-$$

can be established with some difficulty, the fate of the discharged hydroxyl ions is not clear. In all cases metallic oxides are formed, and frequently hydrogen peroxide can be shown to be present. For the conversion of the M—OH into an oxide, two alternative processes have been suggested

$$2MOH \rightarrow MO + M + H_2O \qquad \text{(a)}$$

$$2MOH \rightarrow 2M + H_2O_2 \qquad \text{(b)}$$

together with supplementary reactions involving peroxides such as

$$MO + OH^- + e \rightarrow MOOH$$

$$MO + MOH \rightarrow MOOH$$

Bockris has examined the electrochemical oxidation of ethylene on platinum at 80°C in sulphuric acid. He found that effective utilization of the current was proportional to the pressure of the ethylene, i.e. there is a linear relationship between the pressure and the extent of ethylene adsorption, and that hydroxyl discharge took place directly through the electrode reaction $M + HOH \rightarrow MOH + MH^+$. He has pointed out that, when platinum is replaced by other metallic anodes, as the bond strength D_{M-OH} increases the water discharge reaction is accelerated and eventually the discharge rate becomes so high that this is no longer the rate-determining step. This now becomes the interaction between ethylene and the surface oxide or hydroxide, and in this reaction a weak $D_{M—OH}$ bond is desirable; thus a relatively sharp optimum would be expected in $D_{M—OH}$.

Oxy-redox Catalysis

Oxy-redox systems likewise involve electron transfer: thus the potential of an electrode in a mixture of Fe^{2+} and Fe^{3+} would be given by

$$V = V_0 + \frac{2 \cdot 303RT}{F} \log\left(\frac{a_{Fe^{3+}}}{a_{Fe^{2+}}}\right)$$

The equilibrium involves the electron transfer reaction

$$Fe^{3+} + e \rightleftharpoons Fe^{2+}$$

and the dynamic equilibrium is facilitated by a catalytic metal which serves as an electron-transfer agent in that it accepts electrons from Fe^{2+} adsorbed on its surface and donates it to similarly adsorbed Fe^{3+}. Efficient catalytic metals evidently should have high exchange currents, i.e. large values of i_0. The rate of electron exchange could be evaluated by examining mixtures of two radioisotopic ions of different valence. This can be accomplished much more readily, as Spiro has pointed out, by examining the rate of chemical equilibration of mixtures of two oxy-redox systems, e.g. Fe^{2+}/Fe^{3+} and Ce^{4+}/Ce^{3+} or of reactions such as $Fe^{3+} + H_2 \rightarrow Fe^{2+}$ and $Fe^{2+} + O_2 \rightarrow Fe^{3+}$ in the presence of the catalytic metals, and comparing these rates with those of the cathodic reduction of Fe^{3+} and the anodic oxidation of Fe^{2+}.

Chapter 7

Catalysis in Hydrocarbons

Introduction

Saturated hydrocarbons can undergo three different types of catalytic scission, namely carbonium ion, free radical and carbanion formation. These may be formally described by the equations:

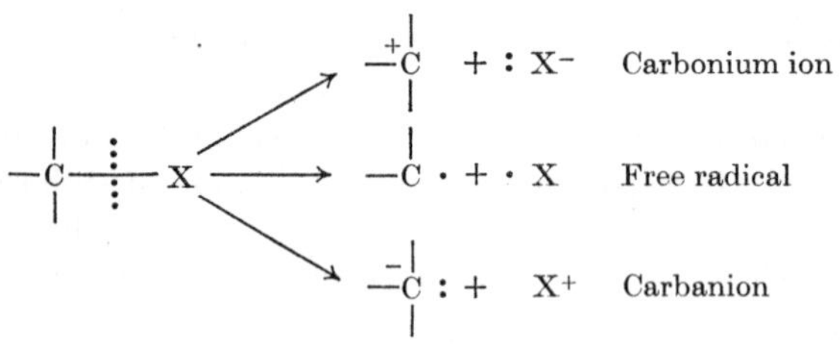

Hydrocarbon free radicals have been identified in a great number of reactions and some, as originally shown by Gomberg, are comparatively stable in the free state. The concept of the carbonium ion was first introduced by Whitmore to assist the interpretation of the mechanism of a number of homogeneous reactions, whereas the idea of the possible intrusion of carbanions in chemical reactions is of more recent date when mechanisms of polymerization with organometallic compounds as catalysts were investigated. Much information as to the reactions of hydrocarbons in the presence of catalysts, especially the transition metals, can be obtained by examining the catalytic exchange reactions of hydrocarbons with deuterium. It was shown some thirty years ago that the hydrogen atoms could be exchanged in hydrocarbons for deuterium when mixtures of the two gases were passed over nickel. A great advance was made by the use of a catalytic chain coupled directly to a mass spectrometer so that the relative proportions of the various deuterium-substituted species could be determined, a method introduced by H. S. Taylor and Turkevich and extensively applied by Kemball.

Hydrocarbon–Deuterium Interactions

The exchange reaction with methane

Methane will exchange with methane-4d, with deuterium oxide and with deuterium itself. All the deuterium-substituted methanes can be prepared on a variety of metallic catalysts, especially those of the transition group; the relative activities being in the order

$$W, Mo, Ta > Rh > Pt(V, Cr, Zr) > Pd > Ni$$

Although all four deuteromethanes are found in the initial products, when the deuterium content of the reacting gases is raised the relative amounts of CH_2D_2 and CHD_3 diminish and the two main products are CH_3D and CD_4. It thus appears that two mechanisms involving C—H bond rupture must be involved. These may be written:

$$\left.\begin{aligned} CH_4 &\rightleftharpoons CH_3 + H \\ CH_3 + D &\rightleftharpoons CH_3D \end{aligned}\right\} \quad (18a)$$

$$\left.\begin{aligned} CH_3 &\rightleftharpoons CH_2 + H \\ CH_2 + D + D &\rightleftharpoons CH_2D_2 \end{aligned}\right\} \quad (18b)$$

The apparent energy of activation for CH_3D formation on nickel is 24 kcal mole^{-1}, appreciably less than those for CH_2D_2, CHD_3 and CD_4 which are 34, 34 and 31 kcal mole^{-1} respectively. The reaction rates are given by:

$$\omega \approx \frac{p_{CH_4}}{p_{D_2}^{1/2}}$$

We can conclude that deuteration proceeds most rapidly by (18a), i.e. through methyl and deuterated methyl groups, and that chemisorbed methylene (18b) is not so readily formed or is more stable *per se*.

Reactions with ethane

Although ethane exchanges with deuterium at an appreciable rate at 110°C on nickel, carbon-bond scission producing CD_4 becomes appreciable only at 150°C. Thus, there is a difference of 50°C between the activation of the C—H and C—C bonds and it is interesting to note that the weaker C—C bond requires the higher temperatures for scission. As is the case for methane, the distribution of deuterium substituted ethanes is not caused by a single process but at least two mechanisms must again be operative, the relative speeds along the two paths varying with the nature of the catalyst. The ratio of the probabilities of the two reactions has been termed the P value by Kemball.

Like methane, ethane can undergo dissociative adsorption on catalytic metals. This takes place in two stages; the half-hydrogenated form (I) being the intermediary:

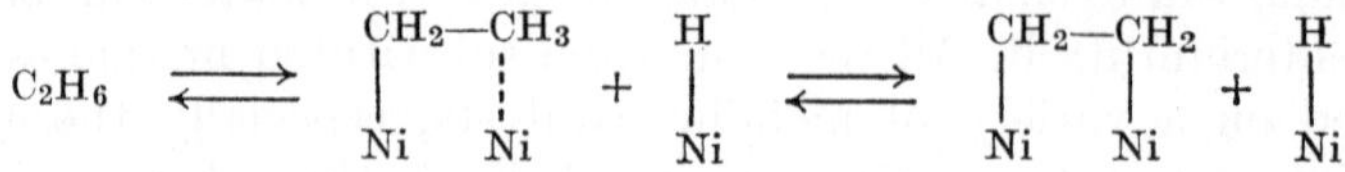

(I) (II)

Deuterium exchange can accordingly take place through either the half-hydrogenated state (I) or through the ethylene adsorbed by two metal atoms (II).

Kemball has evaluated the initial distribution of the various deuterated ethanes to be expected for various values of P, the ratio of the probability of the reaction proceeding through the single point contact reaction (19) to the probability of that proceeding through the two

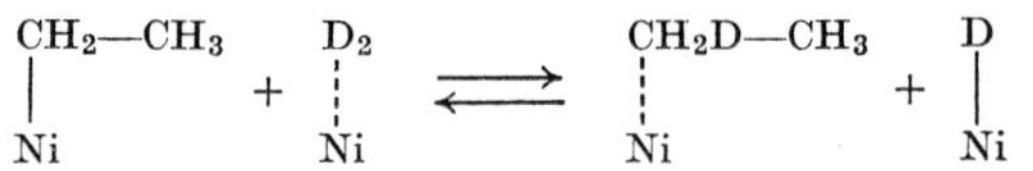

(19)

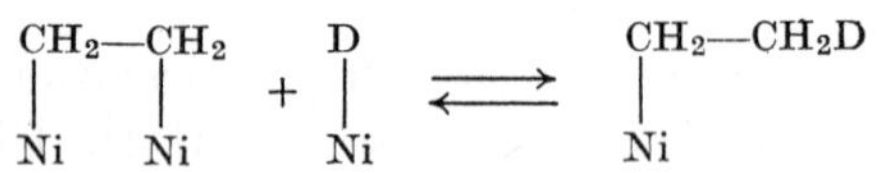

(20)

point contact or ethylene reaction (20). The data in Table XXXVIII indicate that the catalyst has a profound influence on the composition of the deuterated forms in the initial stages and support the view that the reaction proceeds by two independent paths. The reaction velocity of exchange on platinum is roughly $p_{C_2H_6}/p_{D_2}$.

TABLE XXXVIII

Catalyst	P value		Initial distribution					
			d_1	d_2	d_3	d_4	d_5	d_6
Pd	18	Obs.	5	6	8	11	19	51
		Calc.	5·3	7·3	8·7	11·2	16·9	50·6
Mo	0·25	Obs.	82	14	3·0	0·7	—	—
		Calc.	80	17·2	2·5	0·3	—	—
Rh	28	Obs.	4·8	2·9	4·3	7·1	17·1	63·5

Reaction with olefins

The suggestion that one path in the deuteration of ethane is through a chemisorbed species involving two carbon–metal bonds implies that olefin adsorption can take place through the double bond. The double bond in ethylene has two sets of electron pairs: the first order or σ and the second order or π. The σ is a normal C—C bond with axial symmetry and two electrons with antiparallel spin; the C—C distance is 1·547 Å with a dissociation energy $\sim$ 70 kcal mole^{-1} and a characteristic frequency of 1300 cm^{-1}. The two π-electrons also have antiparallel spin and form a weaker bond of 50 kcal mole^{-1}. The π-type can be excited from the singlet ground state in various ways (Fig. 11). The singlet state absorbs at 1800 Å in the ultraviolet and has no bonding orbital. In the N^1 form, the two π-electrons have been moved into antibonding orbitals. An

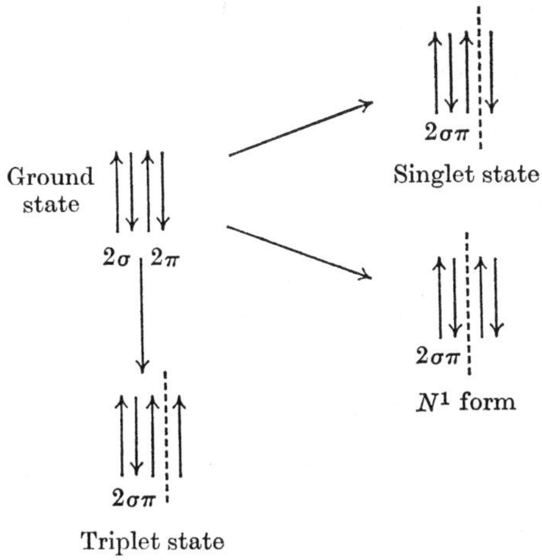

FIG. 11.

ionic state not far from the ground state is thus possible. The triplet state, with antibonding electrons, is an internal radical and the spins of the two π-electrons can be uncoupled in a magnetic field such as with an atom or radical with a free electron.

There are thus three catalytic mechanisms possible with an olefin: a cationic and an anionic one with the N^1 form where E is $\sim$ 23 kcal mole^{-1} with a low frequency factor (both of these are favoured by low temperature) and there is in addition a radical mechanism, with the triplet form where $E \sim 42$ kcal mole^{-1}, proceeding at higher temperatures. A radical catalyst raises the ground state of the olefin non-adiabatically to the excited triplet state and then combines with one of the electrons that have been promoted to antibonding orbitals. The other of the two π-electrons of the original double bond is left unchanged exhibiting the characteristics of a free valency.

An olefin adsorbed on a transition metal catalyst can thus be regarded as π bonded or as a two σ-bonded structure. One course in the deuteration of ethane involves, as we have seen, the conversion of a σ-bonded ethyl radical to a π-bonded or a two σ-bonded olefin. This has been termed the $\alpha\beta$-process whereas that involving only one bond as in the methyl–methyl conversion is called the α or $\alpha\alpha$. We might thus anticipate the deuteration of an olefin proceeds through this type of reaction, i.e. deuterium addition to form a σ-bonded alkyl radical followed by the reforming of the 2σ- or π-bonded now deuterated olefin.

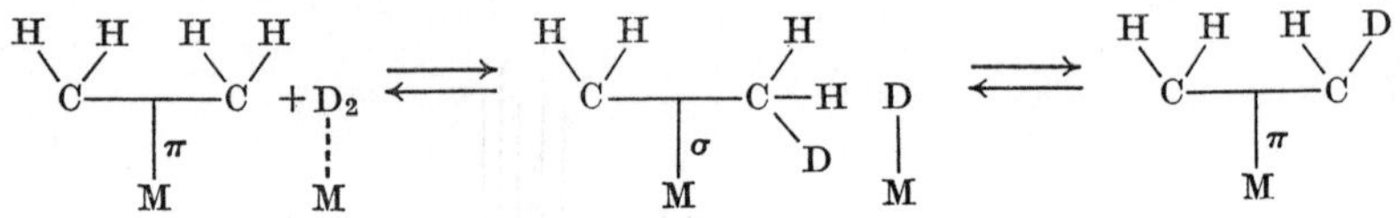

(21)

or

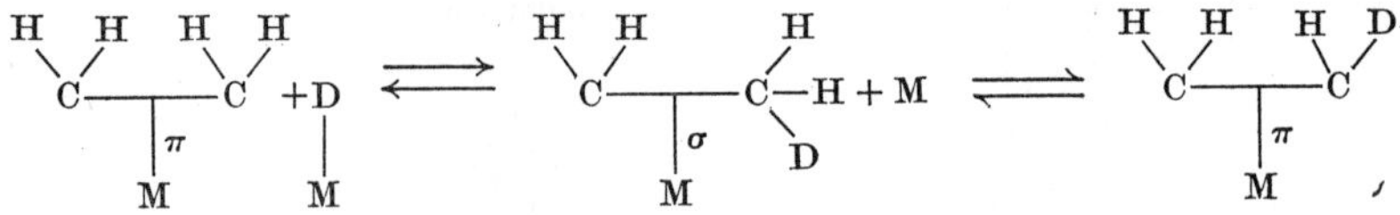

(22)

The strength of the π-bond is difficult to evaluate since electron donation into empty d-orbitals of the metal is in part compensated for by electrons in the d-orbitals of the metal interacting with a vacant antibonding orbital of the olefin. In aromatic molecules at least the net transfer renders the hydrocarbon positive with respect to the metal, even ethylene on nickel has a surface potential of some 0·83 volt. Substitution of alkyl groups for hydrogen in ethylene leads to a reduction in the π-activity of the carbon antibonding orbital, and the result appears to be a weakening of the metal–olefin bond.

Olefins could, like the saturated hydrocarbons, undergo dissociative adsorption on the transition metals. Evidence for this will be considered when we examine the mechanism of hydrogenation of ethylene. The dissociative mechanism can also lead to deuterium exchange:

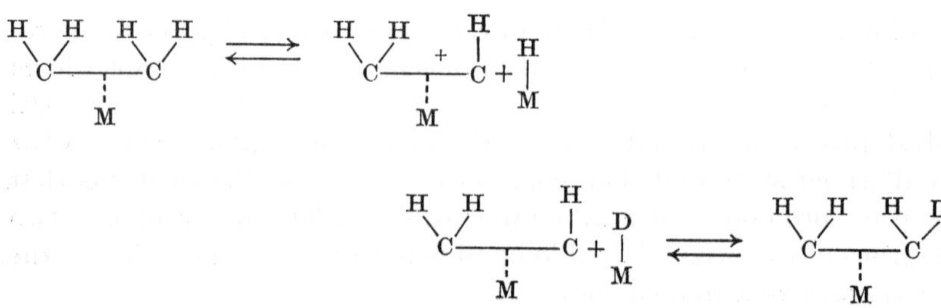

(23)

Reactions with cyclic hydrocarbons

Rooney and Burwell have foundthat in deuterium exchange with cyclic hydrocarbons, e.g. with cyclopentane on palladium all the ten hydrogen atoms both *cis* and *trans* are exchangeable; this is also the case for other cyclic compounds including cyclohexane. Clearly, the $\alpha\beta$ process is ruled out and we must regard these reactions as involving π-bonding only, and instead of metal–carbon ligands we have arene metal complexes. Deuterium exchange in benzene would thus be depicted as follows:

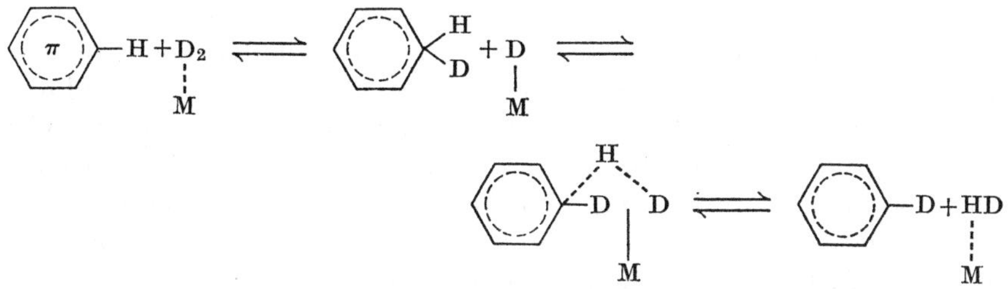

Support is given to the view that π-bonding is involved because the catalytic activities of a nickel catalyst for the hydrogenation of benzene and of substituted benzenes are closely related to the stabilities of their π-complexes (Table XXXIX). The insertion of aliphatic groups in

TABLE XXXIX

Reactant	E (kcal mole^{-1})	log A	Ionization potential (volts)	Stability of the π-complex with: HCl	Picric acid
C_6H_6	14·2	5·8	9·24	0·66	0·83
$C_6H_5CH_3$	13·5	5·4	8·82	1·00	1·00
p-xylene	11·1	4·0	8·49	1·09	1·19
$C_6H_5C_2H_5$	10·4	3·7	8·77	1·15	0·83
Mesitylene	8·0	1·8	8·29	1·73	1·33

the ring raises the electron concentration there, and thus lowers the ionization potential and enhances the electron-donating power; *o*-xylene and cumene are better electron donors than benzene, naphthalene or styrene, and methylstyrene is better still. That the benzene is strongly adsorbed is also noted from the fact that the velocity of hydrogenation is proportional to the hydrogen pressure only, and is independent of the pressure of the benzene, i.e. $\omega = Kp_{H_2}$. There is thus a direct relationship between the ease of π-bonding and the activation energy for hydrogenation. Once the resonance energy of the benzene is lost by the formation of cyclohexadiene, its conversion to cyclohexene and cyclohexane proceeds with far greater velocity. On a platinum catalyst the rate of hydrogenation between 10 and 100 mm Hg and at temperatures between 20 and 100°C, $\omega \propto p_{H_2}$ but the rate of deuterium exchange is independent of the deuterium pressure and $\omega \propto p_{C_6H_6}$ and, in contrast to ethylene, the ortho–para conversion is not inhibited by benzene vapour.

Double bond migration and cis–trans *isomerism*

In addition to the exchange reactions on transition metal catalysts there are two other important reactions of the double bond, namely migration and *cis–trans* isomerism. When deuterium was first employed it was noted that but-1-ene could be converted into but-2-ene and vice versa on nickel catalysts. Double bond migration proceeds to the thermodynamic equilibrium. Measurement of the three reactions, viz. hydrogenation, exchange and double bond migration, on nickel between 75° and 125°C gave for the apparent energies E of activation in kcal mole^{-1}: exchange, 9·0; double-bond migration, 5·9; hydrogenation, 5·6 below 130°C–9·5 above 130°C. These values of E can be compared to the following heats of reactants:

$$CH_2{=\!=}CH_2 + H = C_2H_5 + 37{\cdot}8 \text{ kcal mole}^{-1}$$
$$C_2H_5 + H = C_2H_6 + 97{\cdot}5 \text{ kcal mole}^{-1}$$
$$CH_2{=\!=}CH_2 + H_2 = C_2H_6 + 32{\cdot}6 \text{ kcal mole}^{-1}$$

The kinetics of the double bond migration and exchange reaction were found to be identical at 65°C, viz.

$$\omega = K p_{Bu}^{1/2} p_{H_2}^{1/2}$$

The double bond migration proceeds at low temperatures more rapidly than the exchange reaction, but does not do so on transition metal catalysts in the absence of hydrogen.

We have noted that the exchange mechanism in transition metal catalysts can best be interpreted as proceeding through the half-hydrogenated state (21) or (22) and double-bond migration as well as the *cis–trans* isomerism can be accounted for in a similar manner by hydride hydrogen addition and subtraction.

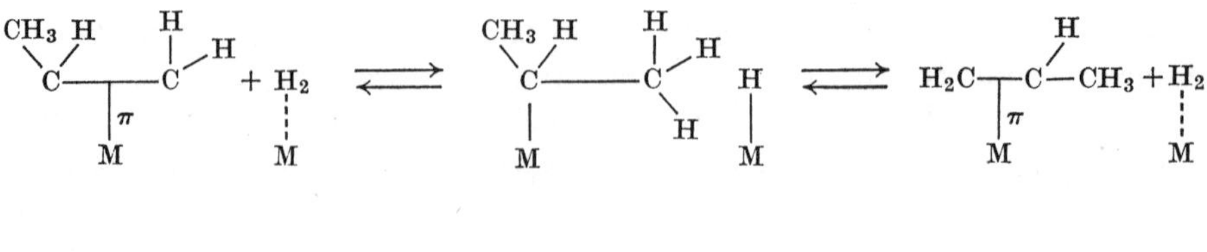

(24)

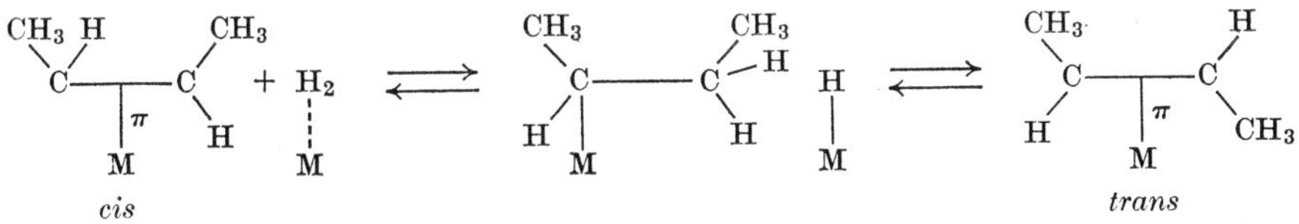

(25)

Hydrogenation of ethylene

In the reactions of olefins with hydrogen at catalytic surfaces deuterium exchange reactions, double-bond shift and *cis–trans* isomerism can take place. Basically, these reactions can proceed in two different ways: proton addition followed by elimination or proton elimination with subsequent reincorporation as in the saturated hydrocarbons. Experimental data, using catalysts both of the transition metals as well as oxides and silico-aluminas, have been interpreted on each of these assumptions and no very convincing case for the correctness of either reaction has yet been made out since they were first investigated some thirty years ago. The evidence on the whole supports the view that, in these reactions, a proton-addition mechanism leading to the half-hydrogenated state before reversion is more common than a proton elimination or a catalytic hydride formation reaction favoured by oxide catalysts, although both reactions take place.

There is yet another reaction to be considered, namely saturation of the double bond, or hydrogenation. Although this is one of the oldest of catalytic reactions and is the basis of a number of important industries, we are confronted with problems similar to those already indicated. This reaction appears to be even more complex and it has frequently been claimed that the hydrogenation process must follow a path different from that leading to the reactions already considered. The hydrogenation reaction takes place readily at low temperatures on a variety of transition metals having *d*-band vacancies and lattice spacings on suitable planes of about 2·6 Å (Ni, 2·47; Pd, 2·75; Ru, 2·65). Table XL shows for nickel the distribution of site pairs. These data indicate that the [111]

TABLE XL

Face	Area (Å^2) containing one 2·47 Å spacing	Area (Å^2) containing one 3·50 Å spacing
[111]	1·76	nil
[110]	8·65	8·65
[100]	3·07	6·05

plane should adsorb ethylene but not acetylene, equal adsorption of the two gases should occur on the [110] plane and on the [100] plane the two gases should be adsorbed in the ratio of 2:1. Also, the [111] plane should have the highest capacity for the adsorption of ethylene.

From the alloys investigated it would appear that ethylene is too strongly adsorbed relative to the hydrogen for optimum reactivity for, if the *d*-band of nickel is progressively filled with valency electrons of copper, the catalytic activity rises until 60% copper is present when the *d*-band is practically full. It is possible that in this reaction ethylene is reacting as a carbanion, the *d*-band being an electron donor, although, as we have noted, on the pure metal ethylene has the positive dipole sign of a carbonium ion.

On silica–alumina catalysts the rate of hydrogenation decreases as the acidity of the catalyst increases and, incidentally, the hydrogen–deuterium exchange reaction behaves in a similar manner. The mechanism of the reaction originally advanced by Polanyi and Horiuti envisages the following steps

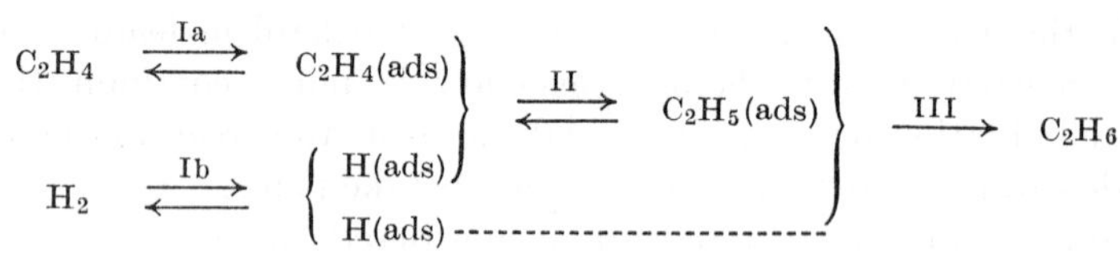

involving the dissociative adsorption of hydrogen and the formation of ethane through the half-hydrogenated state. Evidently, the rate-controlling step might reside in any of the four individual steps that are involved. It may be noted that exchange reactions involve the reversibility of step II as well as of steps Ia and Ib. The hydrogenation of ethylene on nickel is remarkable in that, as Zur Strassen noted, the velocity–temperature curve suffers an inversion at a temperature that depends on the partial pressure of ethylene. Thus, two reactions are involved: one with a positive energy of activation being rate controlling below the critical temperature and one with a negative temperature coefficient above that temperature.

The relationship between the critical conversion temperature and the partial pressure of the ethylene is given by:

$$\frac{1}{T_{crit}} = \frac{1}{5300}(14{\cdot}5 - \log p_{C_2H_4})$$

Horiuti, who has examined the implications of the various steps proposed, concluded that step Ib, i.e. the dissociation of hydrogen, is the controlling reaction below the inversion temperature and that step III takes control above that temperature. If we denote the respective velocities of reaction as ω_{I}, ω_{II}, ω_{III} respectively, then Horiuti's treatment leads to:

$$\frac{1}{\omega} = \frac{1}{\omega_{Ib}} + \frac{1}{\omega_{III}}$$

where

$$\omega_{Ib} = \frac{kT_{n}N_{H_2}}{hf_{H_2}} e^{-E_{Ib}/kT}$$

and

$$\omega_{III} = \frac{kT_{n}N_{C_2H_4}}{hf_{C_2H_4}} e^{-E_{III}/kT}$$

N_{H_2} and $N_{C_2H_4}$ are the concentrations and f_{H_2} and $f_{C_2H_4}$ are the partition functions of molecular hydrogen and ethylene respectively:

$$f_{H_2} = \frac{(2\pi mkT)^{3/2}}{h^3} \frac{4\pi^2 IkT}{h^2}$$

$$f_{C_2H_4} = \frac{(2\pi \bar{m}kT)^{3/2}}{h^3} \frac{2\pi^2(2\pi \bar{I}kT)^{3/2}}{h^3}$$

n is the number of active sites per cm^2, m and $\bar{m}$ the masses and I and $\bar{I}$ the moments of inertia of the respective reacting gases.

Horiuti found $E_{Ib} = 12$ kcal mole^{-1} and $E_{III} = -12$ kcal mole^{-1}, the velocity being proportional to the hydrogen pressure at low temperatures and proportional to $p_{H_2}p_{C_2H_4}$ at high temperatures. We note that in the scheme the relative rate of deuterium exchange in the olefin to the rate of forming a deuteroethane will depend on the fate of the half-hydrogenated complex C_2H_4D. If the chance of ethylene desorption is greater than the chance of hydrogenation the exchange reaction is favoured or the more stable the olefin–metal complex the smaller will be the extent of deuterium substitution relative to hydrogenation.

Table XLI gives Bond's data concerning the chance of desorption to that of hydrogenation. Using a nickel catalyst both hydrogenation and exchange occur at measurable rates between 20 and 200°C. At

TABLE XLI

Ru	Os	Rh	Ni	Pd	Pt	Ir
80	55	25–60	23	25	2	1–5

20°C hydrogenation is faster than exchange and at 155°C the exchange reaction is the more important, as will be seen from the figures in Table XLII. Equality of rates occurs at about 60°C. The apparent energy of

TABLE XLII

	Partial pressures			D content % H_2
	$p_{C_2H_4}$	$p_{C_2H_6}$	p_{H_2}	
Initial	20	0	24	62
Final	13	7	17	22

activation of the exchange reaction is thus greater than that of hydrogenation, being some 6–9 kcal mole^{-1} greater over the temperature range 55–120°C (i.e. $E_{\text{exchange}} = 18$–20 kcal mole^{-1}) and having the low value of $E_{\text{exchange}} = 4$ kcal mole^{-1} on nickel at 200°C).

The proposal for a dissociative mechanism for the hydrogenation process received some support when it was found that a chemisorbed ethylene molecule occupied twice the space occupied by two hydrogen molecules. The deduction that dissociation to an acetylenic compound can take place received further support from infrared analysis of ethylene on platinum suggesting the presence of $\overset{\text{C}}{\underset{\text{M}}{|}}$ $\text{M}\diagup\overset{\text{C}}{}\diagdown\text{M}$ linkages, and on nickel a half-hydrogenated radical $CH_3\text{—}\underset{\underset{\text{M}}{|}}{CH_2}$ as well as an unsaturated residue $\underset{\underset{\text{M}}{\|}}{CH_2}$ have been postulated. Again, if ethylene is admitted to a nickel catalyst in small amounts at a time it is found that ethane is desorbed, indicating that reactions of the type:

$$C_2H_4 \rightarrow C_2H_{2_a} + H_2 \rightarrow C_xH_{y_a} + C_2H_6$$

and

$$C_2H_4 \rightarrow C_2H_{3_a} + H_2 \rightarrow C_2H_{5_a} \qquad 2C_2H_{5_a} \rightarrow C_2H_4 + C_2H_6$$

must be involved.

Information with respect to the acetylenic complex is somewhat at variance. According to some investigations it is relatively inert, building up a polymeric structure –CH—C—C—C– and eventually forming a nickel carbide; other experiments suggest that it can be rehydrogenated back to ethane. This dissociation evidently does not proceed far in the presence of hydrogen, since the activity of the catalyst is well maintained. The mechanism of *cis* addition might be formulated on the dissociative hypothesis by a pathway such as the following in which a Rideal mechanism is involved to effect reaction between gaseous hydrogen and a chemisorbed CH_2—CH radical.

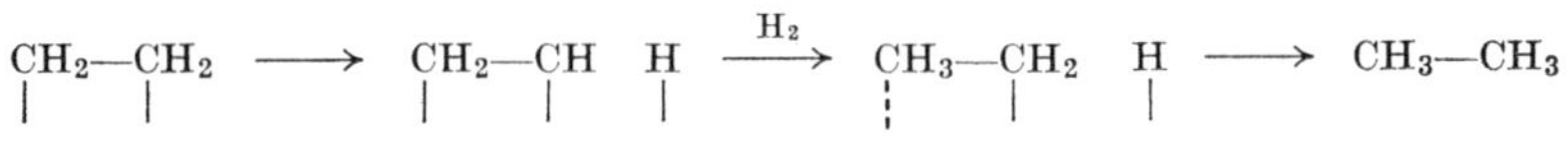

The ortho–para hydrogen conversion is eliminated where ethylene is present in the gaseous phase. Although this is frequently taken as proof that chemisorbed hydrogen is not a constituent of the surface phase, it can also be regarded as indicative of the displacement of molecular hydrogen by the more strongly adsorbed ethylene above the surface phase i.e. in the van der Waals field. The same inhibiting action of ethylene is also noted in the hydrogen–deuterium exchange reaction up to the Zur Strassen temperature of about 150°C. Here, some deuterated ethylene can be detected in the product but this is not conclusive as an indication that some dissociative mechanism must be operative since the hydrogen–deuterium and ortho–para conversion can proceed though a half-hydrogenated state mechanism. Ethylene will react readily with chemisorbed hydrogen, and the suggestion has been made that the mechanism of hydrogenation consists in the chemisorption of hydrogen in the vacancies between the twopoint adsorbed ethylene and its removal by further reaction.

Pressure dependence

According to the mechanism of Polanyi and Horiuti, the rate of reaction below the Zur Strassen inversion temperature should be governed by the rate at which atomic hydrogen (step Ib) is formed at the surface and above that temperature the rate should be governed by the repression of the back reaction C_2H_5(ads) → C_2H_4(ads) and by the supply of atomic hydrogen (step III). The rate should be proportional to the partial pressure of ethylene or

$$\omega \propto p_{H_2} \text{ below } 170°\text{C}$$

$$\omega \propto p_{C_2H_4} p_{H_2} \text{ above } 170°\text{C}$$

Adopting the simplest premise that the ethylene is strongly adsorbed and reacts with chemisorbed hydrogen we obtain for the fraction of the surface covered with ethylene

$$\theta_{C_2H_4} = \frac{p_{C_2H_4}}{1 + p_{C_2H_4}}$$

and for the fraction covered with chemisorbed hydrogen

$$\theta_{H_2} = \frac{p_{H_2}}{1+p_{C_2H_4}}$$

So the reaction velocity should be

$$\omega \propto \theta_{C_2H_4}\theta_{H_2} = \frac{p_{C_2H_4}p_{H_2}}{(1+p_{C_2H_4})^2}$$

if gaseous hydrogen and not the chemisorbed species is involved, then

$$\omega \propto p_{H_2}\theta_{C_2H_4} = \frac{p_{H_2}p_{C_2H_4}}{1+p_{C_2H_4}}$$

which can be distinguished from the former reaction in that there should be independence but no maximum in the curve as the ethylene pressure or temperature is increased.

Early experiments showed that, on copper, at 0–20°C $\omega \propto p_{H_2}/p_{C_2H_4}$ and at 150–300°C $\omega \propto p_{H_2}p_{C_2H_4}$, and on platinum from 0 to 230°C $\omega \propto p_{H_2}/p_{C_2H_4}$. On nickel $\omega \propto p_{H_2}$ when ethylene was in excess, and $\propto p_{C_2H_4}$ when hydrogen was in excess. We have already noted Horiuti's dependencies on the temperature.

The investigations of Schuit and von Reijen on a nickel on silica catalyst revealed that these simple relationships between reaction velocity and partial pressure of the reactants were by no means exact. Thus, for nickel they found $\omega \propto p_{H_2}^{0\cdot67}p_{C_2H_4}^{-0\cdot08}$ between 195 and 273°K with $E_{hydrogenation} = 8\cdot4$ kcal mole^{-1}. In Table XLIII are given the exponents of the partial pressures for a number of catalytic metals.

Although the temperatures chosen were such as to give measurable velocities the variations in the exponents with temperature unfortunately were not recorded. Schuit and von Reijen also made the

TABLE XLIII

Exponents of the partial pressures in $\omega \propto p_{H_2}^m p_{C_2H_4}^n$

Metal	Temperature (°K)	m	n
Rh	197	0·85	−0·79
Ru	203	0·95	−0·59
Co	213	0·56	−0·19
Ni	233	0·67	−0·08
Pt	232	0·77	+0·25
Pd	243	0·66	−0·03
Fe	303	0·69	−0·04
Cu	353	0·69	+0·06

interesting observation that $E_{\text{hydrogenation}} \sim 10{\cdot}7$ kcal mole^{-1} for nearly all their catalysts, and lower values were recorded only when the catalyst was partially poisoned (presumably by poisons that could be removed on elevation of the temperature); on the other hand, the activity factor or the A term in the Arrhenius equation had a thousandfold variation. If both hydrogen and ethylene are reversibly adsorbed, the former as atoms, then we might anticipate that the surface concentration of hydrogen atoms should be some function of $p_{H_2}^{1/2}$ rather than of p_{H_2} and the mechanism proposed by Twigg and Rideal in which the first reaction in the Polanyi Horiuti scheme is replaced by

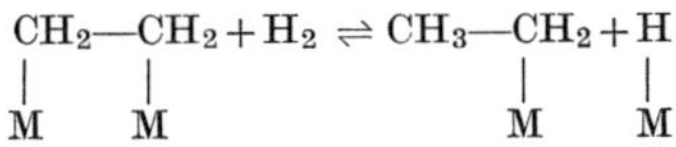

is to be preferred. The fractional exponents in p_{H_2} for different metals may well represent a measure of the relative speeds of the reactions:

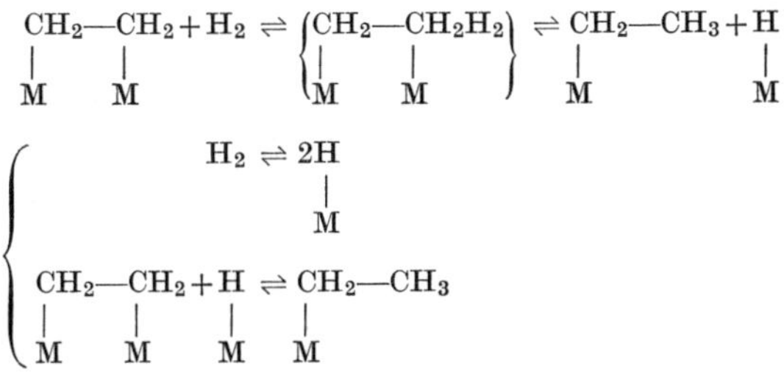

the stronger the olefin metal bond the more stable would be the molecular complex involved in the first scheme.

Merta and Ponec support a dissociative mechanism for the catalytic decomposition of cyclopropane on the transition metals.

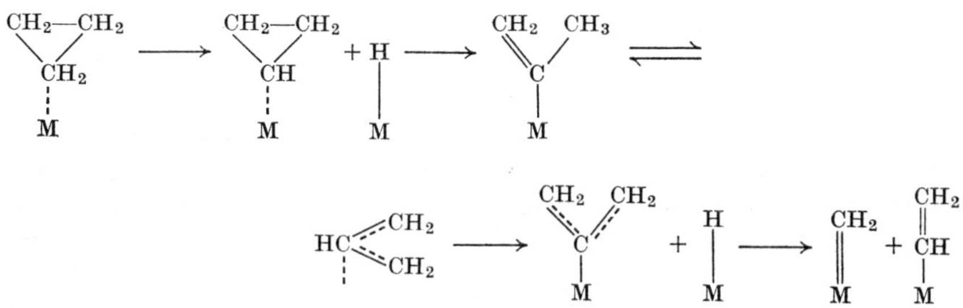

Carbonium Ion Catalysts

Both double-bond migration and *cis–trans* isomerism can also be effected on the so-called "cracking" catalysts of which alumina, silica and the alumino-silicates are representative. These can act both as proton donors, i.e. as Brønsted acids, and, by reason of their electron deficiency, as Lewis acids as well. Whereas in the transition-metal catalysts the energy of activation is dependent on the promotional work involved in electron transfer from electronic bands, in these catalysts the crystal field of a unit of the solid crystalline catalyst is involved. As we shall have occasion to note, catalytic hydrogenation can also be effected by molecules of organometallic complexes where the catalytic activity is determined by the nature and strength of the organic ligand bonded to the central metal ion of a complex molecule. In the majority of these catalysts both Brønsted and Lewis sites are present, as revealed by the infrared absorption of adsorbed NH_3 and NH_4^+ or by dyestuff adsorption. Electron spin resonance measurements of adsorbed ethylene and but-2-ene on alumino-silicate catalysts reveal both a carbonium ion and a complex on a more acidic portion of the substrate. Adsorbed triphenylmethane, however, gives a carbonium ion on a Lewis site.

During the dehydration of hydrated alumina, pairs of surface hydroxyl groups combine to form water, leaving surface oxide and isolated hydroxyl groups. For every molecule of water formed, one oxide group is left in the top layer and one aluminium ion is left in an incomplete octahedral site on the next lower layer. The oxide groups develop on dehydration into domains of regular lattice structure, and the hydroxyl groups tend to concentrate at the boundaries of these domains. Further dehydration results in the production of two or more oxide ions with electron donating properties occupying adjoining surface sites and two or more vacancies in adjoining sites, thus producing an abnormally exposed aluminium ion with its concentrated localized positive charge behaving as a Lewis site.

Alumina requires heating to some 900°C before electron transfer properties are detected and it is claimed by some that electron transfer occurs only when oxygen is present. Thus although weak Lewis sites can form a weak charge transfer complex with an electron rich hydrocarbon, this complex remains spin paired and no colour change is observed, for the carbonium ion is only found where oxygen is present. These two complexes may be represented as

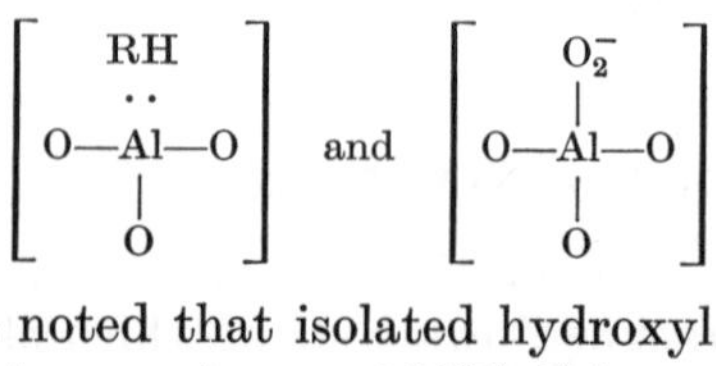

R^+H respectively

It may be noted that isolated hydroxyl groups can be removed as water at high temperatures ~ 900°C; this removal must involve migration of either —OH or a proton leaving a surface free-radical. In a similar manner the surface groups on silica must consist of —Si—OH and Si—O—Si (with bridging O) groups. The removal of water to form the siloxane groups leaves isolated silanol groups on the surface. There should be some 4·5–9·6 silanol groups per 100 Å^2, depending on the crystal facet and the allotropic form of the silica employed. There is evidence to suggest that in the process of dehydration some of the paired silanol groups result in the formation of a zwitterion $—\overset{+}{Si}—O—\overset{O^-}{Si}—$ instead of the more usual siloxane.

The maximum in acidic capacity in the alumino-silicate on the progressive increase in aluminium content is to be found in the range of compounds $\{Si_8(AlO)_{18}\}H$. In low aluminium catalysts the α-crystobalite phase is present changing in turn to bayerite and γ-Al_2O_3 as the alumina content rises.

The origin of the acidity in these catalysts can be ascribed to the replacement of a Si^{IV} lattice ion by an Al^{III} ion; its six-coordination being forced to take on a four-coordinated structure, or the aluminium ion tends to acquire a pair of electrons to fill its *p*-orbital, thus creating a Lewis acid in the absence of water and a Brønsted acid where water is present.

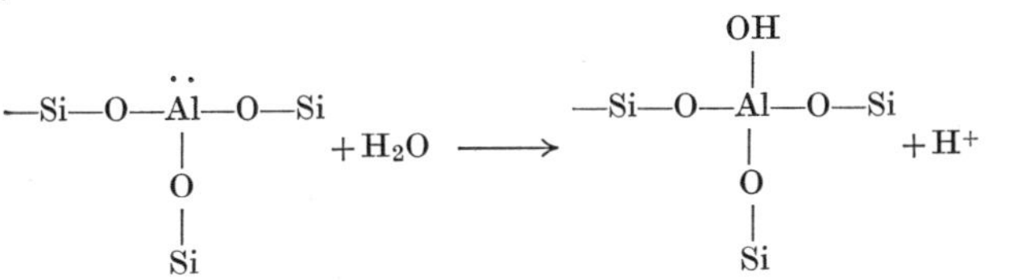

Electron transfer properties develop in these catalysts at temperatures as low as 110°C. Although both Brønsted acids and Lewis acid sites can be identified in this silica–alumina system, the evaluation of their respective contributions to the surface coverage is not easily accomplished. This may be due to the presence of electron-acceptor sites of different strengths which react differently to different electron donors. One of these abnormal groups may be due to the presence of oxygen in a terminal SiO_2 group:

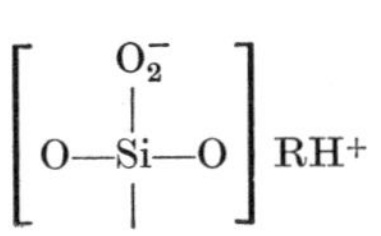

The alumina–silica catalysts can bring about double-bond shift, *cis–trans* isomerization, polymerization of olefins and in addition cracking of saturated hydrocarbons, and alkylation as well as skeletel re-arrangement of a carbon chain. In addition, catalysts in which the replaceable hydrogen has been exchanged for deuterium can in turn exchange with hydrogen in adsorbed hydrocarbons.

As is the case for the transition metal catalysts, two different mechanisms are possible for the production of carbonium ions. The addition reaction to produce the half-hydrogenated state can take place with the Brønsted acids; thus, butene can be converted to the butylcarbonium ion by a Brønsted proton.

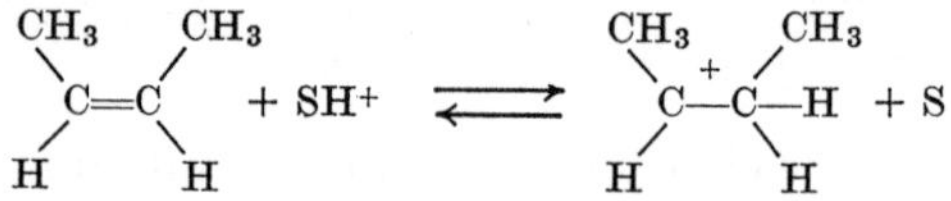

However, we can also envisage the formation of a butenyl carbonium ion by hydride abstraction on a Lewis site, yielding planar *cis-* and *trans-*butenyl ions.

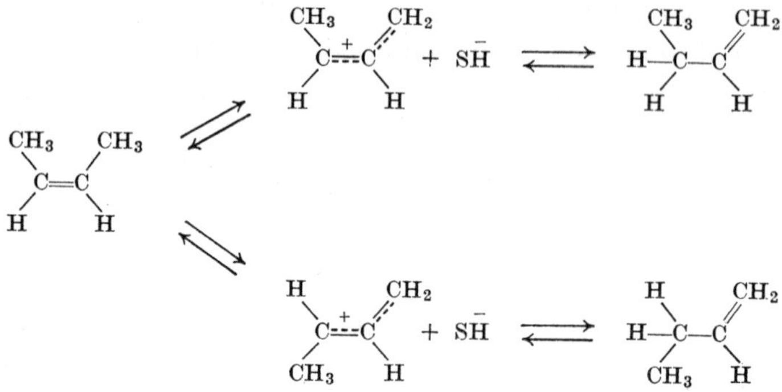

It can be only oxides with strong Lewis acid sites or metals capable of forming hydride hydrogen that would be capable of forming butenyl ions. Butyl ion formation on Brønsted sites would seem to be the more favoured course, a polar adsorption complex and not hydride loss being the extreme of interaction in the former system.

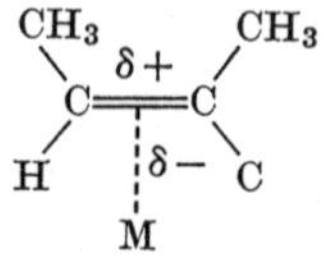

Experiments in which the Lewis sites are inactivated by Na^+ are still active for the majority of these reactions although deuterium exchange appears to involve only Brønsted acids.

In the silica–alumina catalysts the variation in acid strengths, i.e. the hydrocarbon and proton ligands to the metal atom, can be varied by exchanging some of the sodium ions for other multivalent cations, especially those of the rare earths. For this purpose zeolites are usually employed in which the pore structure is sufficiently open to permit access of the hydrocarbon. Little is known at present of the quantitative relationship between the ligand strengths of a catalytic unit in the crystal field, the charge and proximity of the inserted cation but experimentally catalysts some hundred times more active than the original silica–alumina catalysts have been prepared in this way.

Cracking Reactions

With carbonium ion forming catalysts two reactions are possible with the saturated hydrocarbons on Brønsted and Lewis sites respectively.

$$CH_3—CH_2—CH_2—CH_3 + MH \rightarrow CH_3—CH^+—CH_2—CH_3 + H_2 + M$$
$$CH_3—CH_2—CH_2—CH_3 + M \rightarrow CH_3—CH^+—CH_2—CH_3 + MH$$

The reactive carbonium ion may undergo decomposition or react with another saturated hydrocarbon molecule.

In the decomposition of the ion the C—C cleavage takes place one carbon atom distant from the ionic centre, yielding an olefin and a new carbonium ion, e.g.

$$CH_3—CH^+—CH_2—CH_3 \rightarrow CH_3—CH^+—CH_2 \,\vdots\, CH_3 \rightarrow CH_3—CH{=}CH_2 + CH_3^+$$

This has been termed by Whitmore β-scission in which only electron shifts and no atomic or ionic shifts take place.

In the reaction between a carbonium ion and a saturated hydrocarbon a new carbonium ion is also formed, e.g.

$$C_2H_6 + M^+ \rightarrow CH_3—CH_2^+ + MH$$

$$CH_3—CH_2^+ + CH_3CH_2CH_2CH_2CH_2CH_3 \rightarrow$$
$$C_2H_6 + CH_3—CH^+—CH_2CH_2CH_2CH_3$$
$$\rightarrow CH_3—CH{=}CH_2 + CH_2^+CH_2CH_3$$

This primary carbonium ion can go over to a secondary one by electron shift and this can then crack

$$CH_2^+.CH_2CH_3 \rightarrow CH_3CH^+.CH_3$$

Cracking catalysts can also bring about skeletal re-arrangement of the carbon chain to produce iso-paraffins. The following mechanism has been suggested for this re-arrangement.

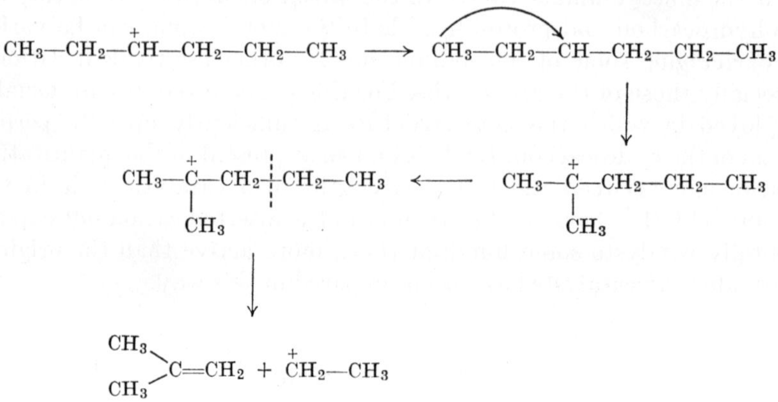

Catalysts capable of bringing about hydrocarbon isomerization at relatively low temperatures may be prepared from platinum containing alumina which has reacted with carbon tetrachloride. The reaction is believed to take place on localized dual sites, i.e. a Lewis acid site and an adjacent metal atom. In metals the order of effectiveness was found to be Pt > Pd > Rh ≫ Ir, Ru; the replacement of a hydroxyl by a chlorine renders the remaining protons more acidic. The dual site arises from the following surface reactions

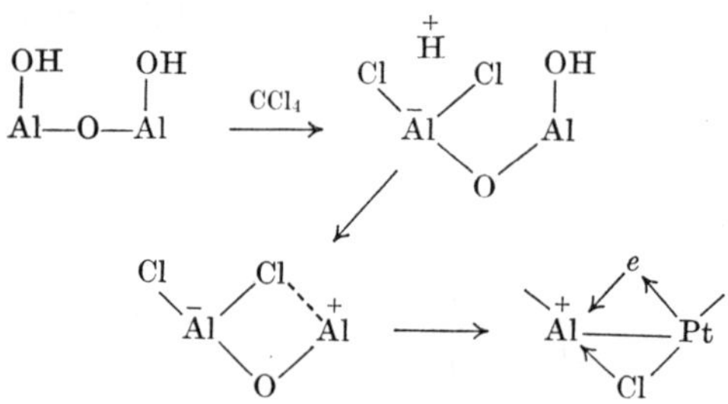

the isomerization taking place through an intermediary positive paraffin ion. Thus, the reforming of methylcyclopentane to benzene can be regarded as taking place by the following steps.

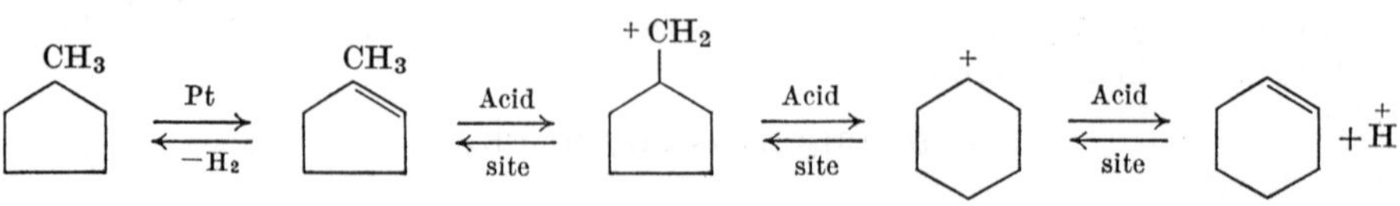

Chromium Oxide Catalysts

Chromium oxide supported on chromia, alumina or silica is found to be an effective catalyst for a number of reaction such as the deuterium exchange with alkanes and benzene, dehydrogenation of cyclohexane, hydrogenation of olefins as well as the polymerization of ethylene. The exchange reactions of the olefins and alkanes with deuterium differ from the same reactions on catalytic metals in that the exchange is by no means as extensive, monosubstituted products being most abundant. The polymerization of ethylene takes place at low temperatures and is found to be of zero order up to 0°C. There is little doubt that there are two active species in chromia, one promoting the polymerization and one bringing about the hydrogen dehydrogenating reactions. This fact accounts for the wide variation in the results obtained for different chromium oxide catalysts.

Of the various oxides (Cr^{II}, Cr^{III}, Cr^{IV}, Cr^{V}, Cr^{VI}) that may be present the experimental work supports the view that only two of the oxides are catalytically significant, the Cr^{V} ion being that necessary for polymerization and the reduced Cr^{II} ion for the hydrogenation and dehydrogenation reactions. The presence of the Cr^{V} ion has been detected by e.s.r. measurements. The signal is weakened on exposure to ethylene but is strengthened if the ethylene–Cr^{V} complex be exposed to water vapour which at the same time cuts down the catalytic activity. Water vapour *per se* has no influence on the signal strength. Gentle heating of a Cr^{VI}-containing catalyst in ethylene results in the reduction of the Cr^{VI} to Cr^{V} and enhances the catalytic activity for polymerization. These observations support the view that polymerization is initiated through the formation of a π-complex of ethylene with a coordinated Cr^{V} complex. If the Cr^{V} complex oxide is regarded as octahedral it will have two empty coordination sites and the complex formation with ethylene and subsequent *cis* insertion of another ethylene provides a ligand mechanism analogous to that of the organometallic catalysts.

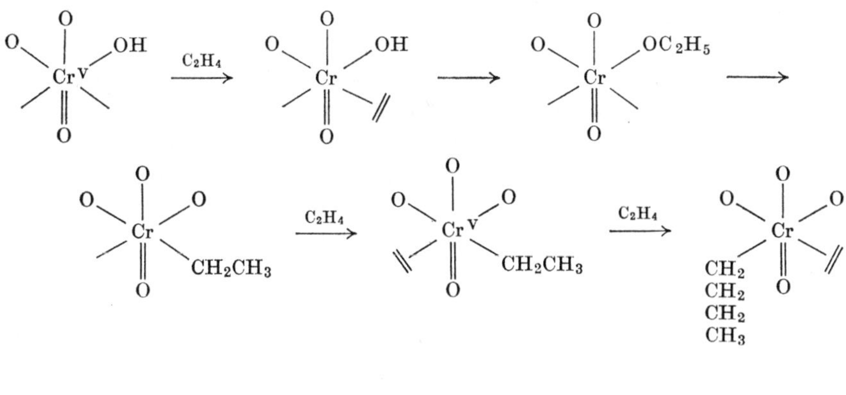

It has also been suggested that the Cr^V complex is not octahedral, but square pyramidal in coordination. In this case there would be but one empty coordination site:

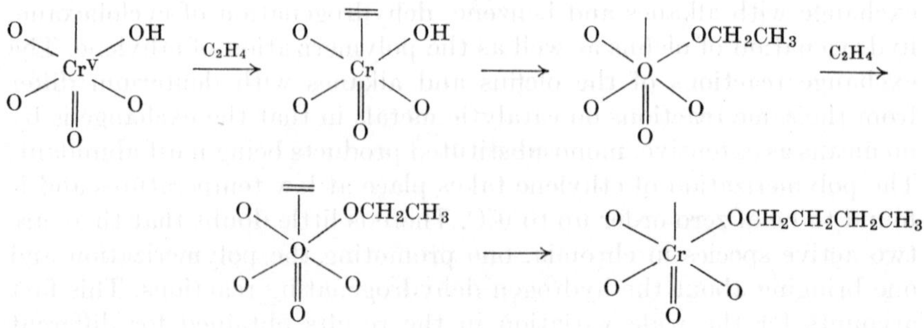

We may add that a third mechanism has been proposed where the Cr^V catalyst initiates polymerization through carbanion growth centres

$$Cr^V + C_2H_4 \rightarrow Cr^{VI}C_2H_4^-$$

On further reduction of the chromium oxide catalyst the presence of both blue Cr^{II} and green Cr^{III} ions can be detected. The ratio of divalent to trivalent ions is determined to a great extent by the equilibrium:

$$Cr^{III} + H_2 \rightleftharpoons Cr^{II} + H_2O$$

although some of the Cr^{III}, present in all probability in solid solution (when in γ-Al_2O_3), is reduced only with great difficulty. The equilibrium of the above reaction is far to the left since no appreciable reduction of Cr^{III} results with hydrogen when saturated with water vapour.

Cr_2O_3 in an oxidizing atmosphere is a p-type semiconductor and partial conversion of the Cr^{III} to Cr^{VI} takes place. It requires heating to some 400°C in a reducing atmosphere to convert it to an n-type e.g. reduction by hydrogen of Cr^{III} to Cr^{II}.

Carbon monoxide but not hydrogen is strongly chemisorbed by Cr^{II} but not by the higher oxides. The heat of adsorption of carbon monoxide is given by Sachtler as from 27 kcal mole^{-1} at $\theta = 0{\cdot}04$ to 13 kcal mole^{-1} at $\theta = 0{\cdot}5$ He has also examined the dehydrogenation of cyclohexane on a Cr^{II} catalyst and found that it proceeds smoothly to benzene at 450°C with an activation energy $E = 26$ kcal mole^{-1} and at a rate proportional to the pressure between 16 and 65 mm Hg; i.e. the rate is controlled by the rate of dissociative adsorption of the cyclohexane where the reaction step is presumably

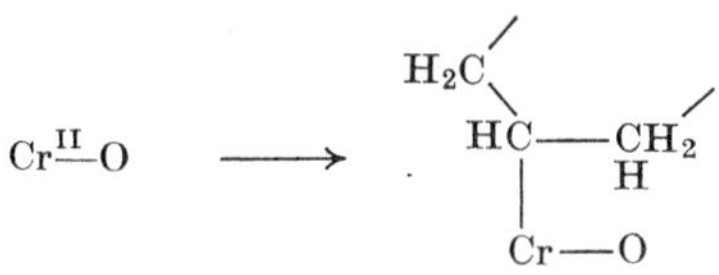

The number of Cr^{II} sites on the catalyst was determined by means of the oxygen uptake and the rate was found equivalent to $1{\cdot}4 \times 10^{-2}$ molecules site^{-1} sec^{-1}. The temperature rate constant in the Arrhenius equation was calculated to be $10^{7 \cdot 3}$ in good agreement with the value derived from the transition state equation on the assumption that both translational and rotational degrees of freedom are lost in the activated complex.

The reverse reaction, namely the hydrogenation of hexene, takes place on Cr^{II} at temperatures as low as 43°C. It is significant that benzene itself is more difficult to hydrogenate but isotopic exchange with deuterium is observable at 80°C.

Carbanion Catalysis

These anionic or base-catalysed reactions are not so clearly defined as the carbonium ion reactions. They are of importance in polymerization processes involving vinyl derivatives. The earliest example of such is probably Tilden's synthesis of rubber from isoprene in the presence of metallic sodium. The alkali metals can combine with bases to form ionic salts such as $K^+NH_2^-$, Na^+naphthyl$^-$, $K^+CPh_3^-$ and $Na^+C_2H_5^-$. The organometallic compound of the coordinate type must have the carbon–metal bond polarized so that the carbon moiety is partly an organic ion. This is the case in the Alfin or Ziegler and Natta polymerization catalysts: the latter must be able to make complexes with halides of the transition metals. Active catalysts can also be prepared by the deposition of sodium or lithium on alumina, but it has not been shown as yet whether these react with hydrocarbons to form carbanions. Carbanion catalysts can react with olefins in three different ways: (a) by abstraction of a proton

$$Na^+B^- + CH_3{-}CH_2{-}CH{=}CH_2 \rightarrow {-}^-CH_2{-}CH_2{-}CH{=}CH_2 + HB$$

here the carbanion is stabilized by resonance

$${-}^-CH_2{-}CH_2{-}CH{=}CH_2 \rightleftharpoons CH_2{=}CH{-}CH_2{-}CH_2^-{-}$$

(b) by direct addition of the anion

$$Na^+Ph^- + CH_2{=}CH_2 \rightarrow PhCH_2{-}CH_2^-{-}$$

This is in contrast to a hypothetical carbonium ion reaction

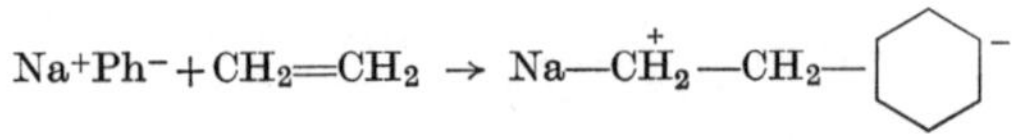

In general, carbon–carbon bonds are stronger than metal–carbon ligands. Thus

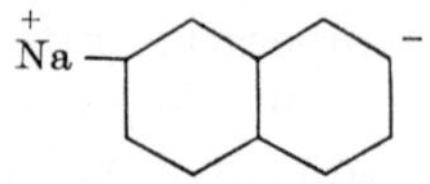

polymerizes ethylene oxide

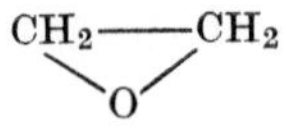

by forming initially

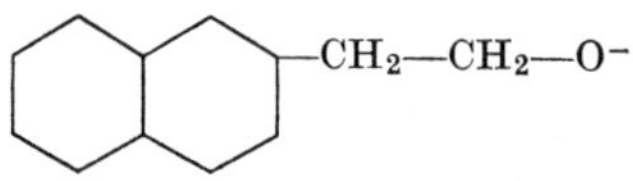

(c) If the donor has a low ionization potential and the monomer has a sufficiently high electron affinity, direct transfer of an electron takes place.

$$D' + M \rightarrow D + M'$$

Thus

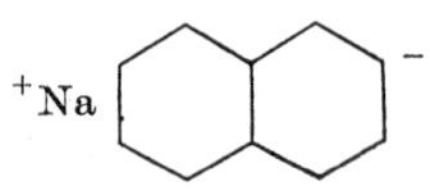

polymerizes styrene by the transfer process

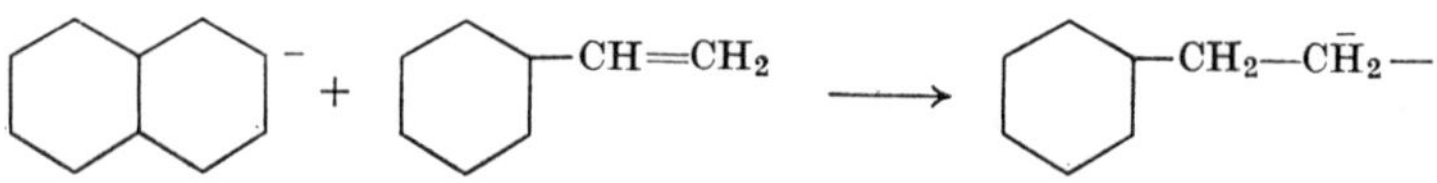

The carbanion is presumably less reactive than a free radical. It is evident that the governing factors which are involved in carbanion formation are the electron affinities and ionization potentials of the catalyst, the solvent, and the monomer of the olefin undergoing polymerization. We find that the formation of the sodium naphthyl catalyst possible in the solvent tetrahydrofuran but not in hydrocarbons which have greater electron affinities.

In the initiation of styrene polymerization it is found, that OH^- ions are ineffective, that NH_2^- is a weak initiator, but $Ph—CH_2^-$ ions are very efficient. Again, it is found that for lithium, $Li—CHX{=}CH_2 \rightarrow Li^+ + CH\overline{X}{=}CH_2$ will polymerize styrene but $Li^+butyl^- \rightarrow Li^+ + butyl—CHX—CH_2^-$ will not polymerize styrene but only methylmethacrylate.

On the addition of another molecule of the olefin to a carbanion we obtain

$$—CH_2—CH_2^- + CH_2{=}CH_2 \longrightarrow \underset{\text{carbanion end}}{^-CH_2—CH_2}—\underset{\text{radical end}}{CH_2—CH_2—}$$

so both carbanionic as well as radical polymerization can take place. As well as the addition reaction we may also get an olefinic re-arrangement

$$CHX{=}CH—CH_2^- + CH_2X—CH{=}CH_2 \longrightarrow CHX{=}CH—CH_3 + {^-CHX}—CH{=}CH_2$$

The acid-catalysed reaction of an aromatic hydrocarbon with a mono-olefin results in the addition of the alkyl groups only to the ring

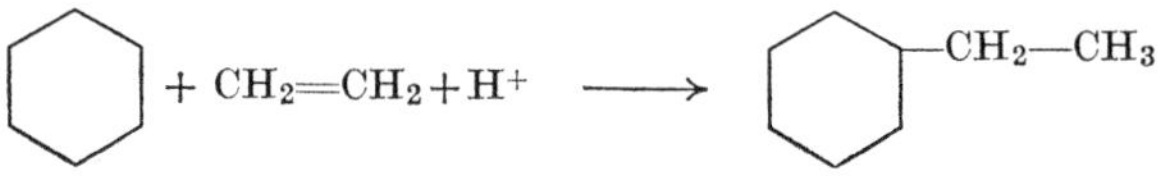

but the base catalysed reaction results in alkylation of the side chain

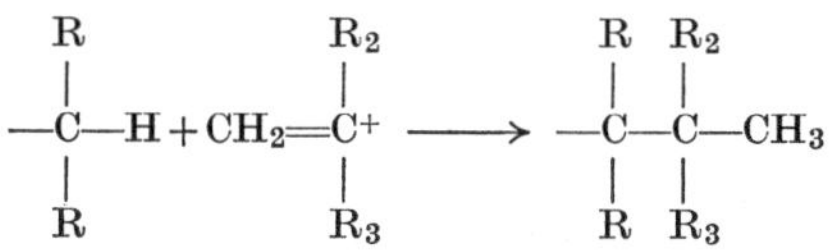

It may be noted that potassium deposited on graphite is an active catalyst for this purpose.

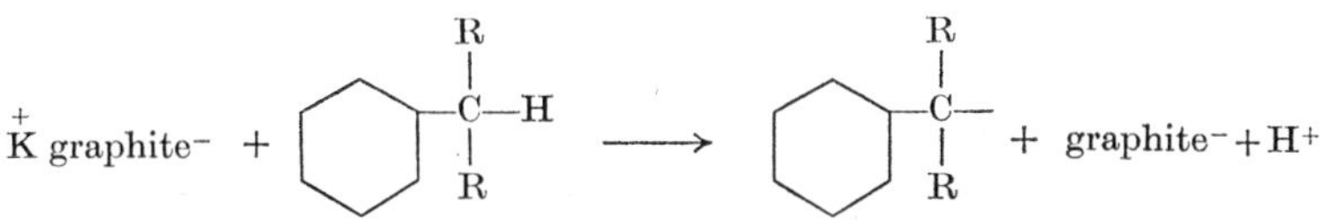

The electron affinity of the monomer plays an important part in the polymerization process. Thus, in a styrene–isoprene copolymer, styrene is less electronegative than isoprene. Consequently, metals of increasingly electropositive character or solvents of increasing basicity depress the isoprene content of a copolymer. Likewise, the more ionic or polarized the initial system, the greater is the *trans*-1,4 and -3,4 content of the finished polymer and the smaller is the fraction of the *cis*-1,4. In the anionic forming catalysts such as $Na^+C_2H_5^-$ we must note that the metal's ionic radius must not be too big and it must be sufficiently electropositive to be able to form the carbanion. The growth proceeds in the form of helical coils. Such polymeric helices tend to get extended in good solvents and compressed in bad solvents. Since *cis–trans* conversion can take place, the helices tend to be more stable in bad rather than in good solvents. If the helix grows more rapidly than it can randomize then we obtain a syndiatactic polymer. Thus catalyst and solvent play an important part in the form of polymer obtained.

$K^+C(Ph)_3^-$ polymerizes styrene to an amorphous powder in benzene, a good solvent, but to an isotactic one in an aliphatic hydrocarbon, a bad solvent, whereas $Na^+C(Ph)_3^-$ produces amorphous polymers in both solvents. The potassium salt produces helices the growth of which is quick enough not to randomize in an aliphatic hydrocarbon.

The older and well-established free radical mechanism is frequently employed in polymerization reactions; thus benzoyl perbenzoate undergoes a bimolecular decomposition in solution

$$C_6H_5OCOOCOC_6H_5 \rightarrow C_6H_5COO\cdot$$

This radical can decompose further to yield phenyl radicals

$$C_6H_5COO\cdot \rightarrow C_6H_5\cdot + CO_2$$

both the phenyl and the benzoyl radicals can initiate chain formation in styrene.

The phenylstyrene complex cannot be broken by hydrolysis whereas the benzoylstyrene complex is readily decomposed. The relative rates of initiation of the two chain systems can be determined by isotopic labelling.

Organometallic Catalysts

In the search for electron transfer mechanisms operative in catalytic processes we have noted that these have included firstly the cooperative electron band concept, then the idea of a more localized catalytic entity involving interaction between an atom or ion and its nearest neighbours. Investigations on the catalytic properties of oxides and salts have, in turn, introduced the concept of a catalytic molecular unit in a crystal field. We have seen that many carbanion reactions can be effected by means of simple organometallic compounds such as the alkali metal hydrocarbon salts. These may be regarded as the progenitors of catalysts composed of coordination compounds of ions of the transition metals.

Here the catalytic unit is the ion which has a suitably disposed ligand system. We are indebted especially to Calvin, to Halpern and to Wilkinson for many of the developments in this field. Not only do they hold out promise of technical development but the analogies with the prosthetic groups of the biocatalysts, the enzymes, are closer than with any other catalytic system.

A number of transition metal ion complexes can catalyse reactions involving molecular hydrogen by their capacity to form reactive metal hydride complexes. These may be formed in three different ways, either by heterolytic or homolytic splitting or by an insertion reaction to form a dihydride. Halpern gives the following examples for hexa-, penta- and tetracoordinate complexes respectively.

$$\mathrm{Ru^{III}Cl_6^{3-} + H_2 \rightleftharpoons Ru^{III}HCl_5^{3-} + H^+ + Cl^-}$$

$$\mathrm{2Co^{II}(CN)_5^{3-} + H_2 \rightleftharpoons 2Co^{III}H(CN)_5^{3-}}$$

$$\mathrm{Ir^{I}Cl(CO)(PPh_3)_2 + H_2 \rightleftharpoons Ir^{III}H_2Cl(CO)P(Ph_3)_2}$$

We note that the heterolytic formation of the hydride in the hexa-coordinated salt possessing six non-binding electrons is a process involving an exchange of a chloride ligand for a hydride ligand, but in the other two processes, which have seven and eight non-bonding electrons respectively, oxidation of the metallic ion is involved. Thus the tendency towards oxidation of the metal is an important factor in the reactivity of these insertion reactions.

Since splitting of the hydrogen molecule is effected in hydride formation by both the heterolytic and homolytic processes, it is possible to bring about the ortho–para hydrogen and the hydrogen–deuterium exchange reactions and in addition proton–deuterium exchange with the hexa-coordinated complex permits deuterium substitution in ionizable hydrogen containing substances such as water. The rates of these reactions are governed by the rate of formation of the hydride complex, which is very reactive.

In recent years the reactions of olefins and the hydride complexes have been investigated.

The hydrogenation of a double bond could proceed by two different methods. Denoting the coordinated salt by S, hydrogenation may proceed through the half-hydrogenated state by interaction with a hetero- or homolytically produced hydride.

$\mathrm{S + H_2 \rightarrow SH + H^+}$			$\mathrm{2S + H_2 \rightarrow 2SH}$
SH + olefin → SH olefin		or	SH + olefin → SH olefin
SH olefin + $\mathrm{H^+}$ → S + paraffin			SH olefin + SH → S + paraffin

or by the alternative path, by the direct addition of hydrogen from the dihydride

$$\begin{aligned} \mathrm{S}+\mathrm{H_2} &\rightarrow \mathrm{SH_2} \\ \mathrm{SH_2}+\text{olefin} &\rightarrow \mathrm{SH_2}\ \text{olefin} \\ \mathrm{SH_2}\ \text{olefin} &\rightarrow \mathrm{S}+\text{paraffin} \end{aligned}$$

It must be noted that as well as the reaction of the olefin with a hydride there are alternative reactions involving the π-bonded olefin–metal complex as the reactant with hydrogen and forming the metal–olefin–"hydride" complex, by homo- or heterolytic splitting or by direct addition.

$$\mathrm{S}+\text{olefin} \rightarrow \mathrm{S}\ \text{olefin}$$

$$\mathrm{S}\ \text{olefin}+\mathrm{H_2} \longrightarrow \mathrm{S}\ \text{olefin}\ \mathrm{H} \xrightarrow{+\mathrm{H}\text{ or }\mathrm{H^+}} \mathrm{S}\ \text{olefin}\ \mathrm{H_2} \longrightarrow \mathrm{S}+\text{paraffin}$$

Whether the reaction proceeds through the hydride or the π-bonded olefin complex will depend on the relative stabilities of the π-bond and the hydride ligand.

Halpern has found that the hydrogenation of fumaric acid in aqueous solution proceeds by the heterolytic splitting of hydrogen by the π-bonded olefin complex of ruthenium chloride, whereas in the hydrogenation of butadiene in aqueous solution by $Co(CN)_5^{3-}$ the reaction according to Kwiatek proceeds through the hydride produced by the homolytic process reacting with the olefin. Wilkinson, whose work will now be discussed in more detail, found that hydrogenation with a rhodium complex proceeded most readily through the metal *cis*-dihydride complex reacting with the olefin.

Although the square complexes $Ir(PPh_3)_2Cl(CO)$ and $Ir(PPh_3)_3H(CO)$ can hydrogenate both ethylene and acetylene, a rhodium complex containing strong π-bond acceptor ligands was found by Wilkinson to be the most active catalyst of this type. The tristriphenylphosphine halides dissociate in solution to give bistriphenylphosphine derivatives with a coordinate ligand vacancy,

$$\mathrm{Rh(PPh_3)_3Cl} \rightleftharpoons \mathrm{Rh(PPh_3)_2Cl}+\mathrm{PPh_3}$$

The bis derivative readily reacts with carbon monoxide and with olefins to give square complexes with *trans* triphenylphosphine groups

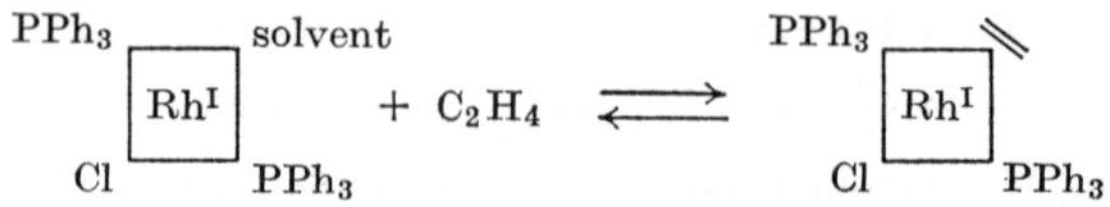

In contrast to the (CO) complex, which is so stable that the carbon monoxide can actually be abstracted from aldehydes, the ethylene complex is readily reversible, exchange between free and coordinated ethylene proceeding even at $-50°C$ with an equilibrium constant of $K \sim 100$ mole^{-1} l. For propylene the constant K is some two thousand times smaller. The decrease in ethylene complex-forming ability $Cl > Br > I$ suggests that the halide is in the *trans* position.

The bis derivative also reacts with hydrogen to form a dissociable dihydride

$$Rh(PPh_3)_2Cl + H_2 \rightleftharpoons Rh(PPh_3)_2ClH_2$$

in a manner similar to the action of the iridium complex $Ir(PPh_3)_2Cl(CO)$. The n.m.r. spectrum and other physical properties of the dihydride indicate that the complex is no longer planar but is now an octahedral *cis* dihydride with one vacant coordination site.

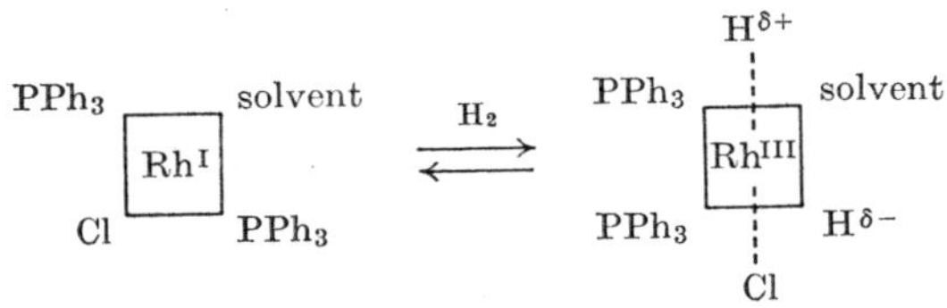

When the vacant coordination site is occupied by an olefin the complex can then form the transition state complex for hydrogenation by *cis* hydrogen transfer

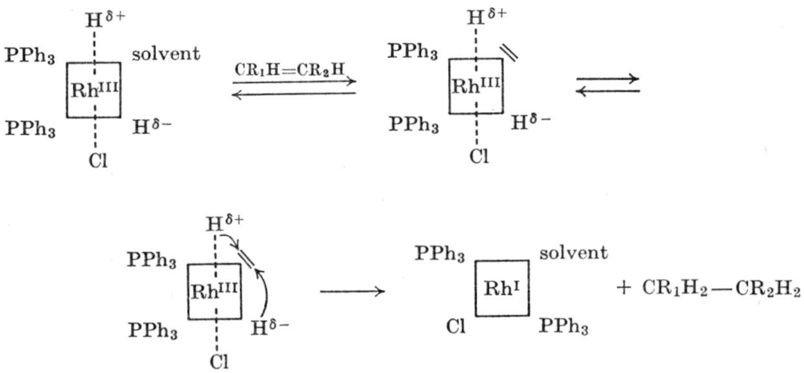

This sequence of events is made possible because of the ease with which the tris derivative can dissociate due to the bulkiness of the triphenylphosphine groups, thus providing the necessary vacant coordination site in the dihydride derivative. Equally important from the point of

view of catalytic activity is the low promotional energy $Rh^{I} \rightarrow Rh^{III}$ and ease of conversion of the Rh^{I} complex into the octahedral Rh^{III} dihydro complex. This energy is naturally increased if the square complex is rendered more stable by a strong π- or 2σ-acid-ligand as with CO. This ligand is weaker for the iridium complex than for rhodium where the promotional energy $Ir^{I} \rightarrow Ir^{III}$ is smaller; the Ir—H bond is also stronger than the Rh—H bond.

Olefins that have relatively small dissociation constants for their π-ligand complexes are readily hydrogenated, but ethylene itself has a sufficiently high π-acidity in the antibonding orbitals to preclude hydrogenation except under high pressures of hydrogen. Ethylene thus acts as a poison for the hydrogen activation step even though transfer from the dihydrido species to ethylene *can* occur. Wilkinson, using this catalyst, found no evidence for *cis–trans* isomerism or double-bond shift, and the hydrogen addition was exclusively *cis*. As would have been anticipated, the bromide and iodide, which can cause progressive weakening of metal–olefin bonds in the complex which activates molecular hydrogen, as well as the M—H bonding by *trans* effects, are more effective catalysts than the chloride. We note that hydrogen addition under these conditions is essentially molecular; there is no dissociation to form a half-hydrogenated state. This receives confirmation in that there is no HD formation with a hydrogen–deuterium mixture; also, with hex-1-ene in such a reaction, a mixture only of dihydrido- and dideutero-hexane was obtained, no paraffin containing only one deuterium being detectable. Wilkinson found the isotope effect in the rates of hydrogenation to be small and less than unity $\omega_{H_2}/\omega_{D_2} = 0{\cdot}9$. Zero point energy is thus actually gained in the conversion of the reactants to the transition state, or the energy lost by bonding and lengthening the Rh—H bond is more than compensated for by the formation of a C—H bond, the zero point energy of which is 1·43 kcal mole^{-1} and is greater than that of the Rh—H bond

$$\Delta E[(\text{Rh—H}) - (\text{Rh—D})] = 0{\cdot}86 \text{ kcal mole}^{-1}$$

In his studies of the kinetics of the hydrogenation based on the equilibria

$$S + H_2 \underset{}{\overset{K_1}{\rightleftarrows}} SH_2(\text{S dihydride})$$

$$S + \text{olefin} \underset{}{\overset{K_2}{\rightleftarrows}} (\text{S olefin})$$

he assumed that these equilibria were rapidly established and maintained during the reaction. On this assumption, the complete reaction cycle can be written

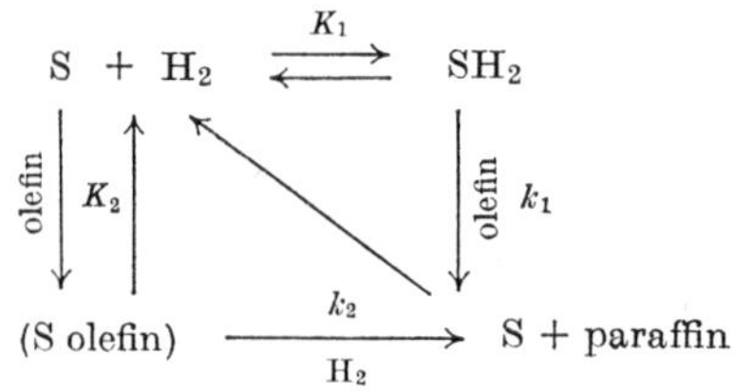

The rate of reaction is then given by

$$\omega = \frac{[k_1 K_1 + k_2 K_2][H_2][S][\text{olefin}]}{1 + K_1[H_2] + K_2[\text{olefin}]}$$

Since the life of the dihydro complex is long and the olefin complex is relatively unreactive, the favoured path of the reaction must be through the dihydride and not through the olefin complex. If we regard the latter reaction as negligible $k_2 = 0$ or

$$\omega = \frac{k_1 K_1[H_2][S][\text{olefin}]}{1 + K_1[H_2] + K_2[\text{olefin}]}$$

which expression was found applicable to his experimental data. Olefins for which K_1 is large and K_2 small are hydrogenated most rapidly; thus, unsubstituted ethylene with a strong π-bond and consequently a large value of K_2 is hydrogenated relatively slowly unless the pressure of hydrogen is high.

For cyclohexene the activation energy was found to be $\Delta E = 22{\cdot}9$ kcal mole^{-1} and the entropy change $\Delta S = 18{\cdot}3$ e.u.

The catalytic hydrogenation of ethylene can also be effected with the aid of Zeise's salt, $K[PtCl_2(C_2H_4)]Cl$. The actual mechanism of the reactions has not as yet been established, but it would appear that molecular addition in the *cis* position takes place in a manner analogous to that for the rhodium complex.

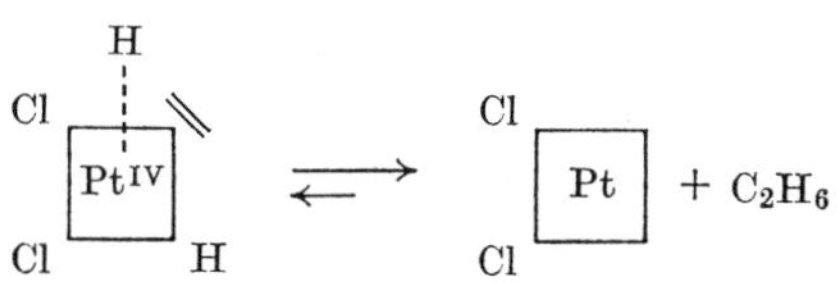

Whereas the hydrogenation of ethylene involves the direct addition of two atoms of hydrogen, it appears that ligand catalysts that bring about the double-bond shift or the polymerization of olefins do so through the intermediary of the half-hydrogenated state. Thus, Cramer has proposed the following sequence for isomerization by means of alcoholic hydrochloric acid solutions of rhodium chloride.

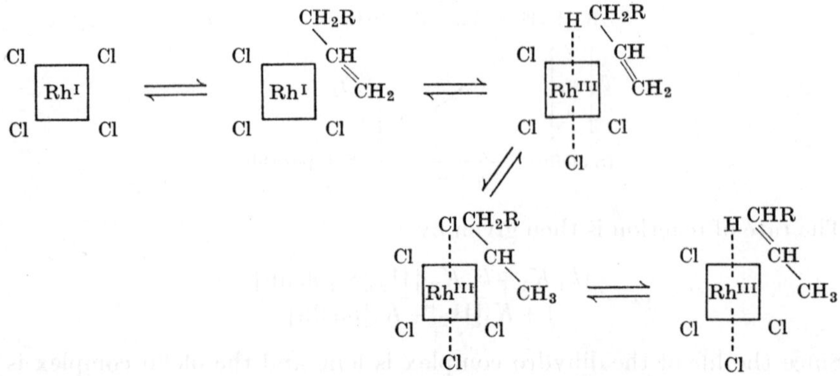

He treats the dimerization of ethylene on the same salt in a similar manner

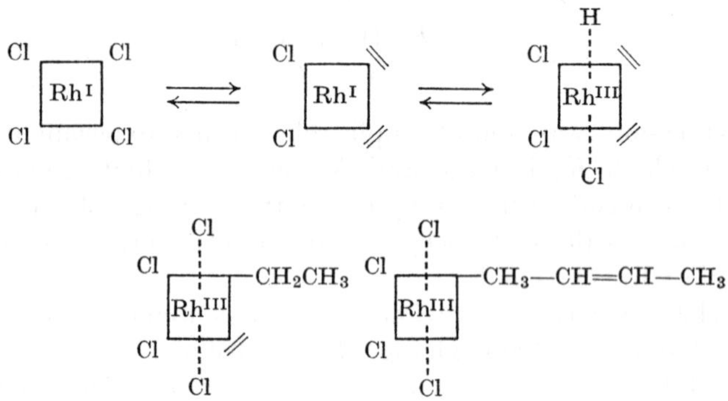

The brown complex resulting from the addition of but-1-ene to a solution of palladous chloride gives all three isomers. It is unknown whether the isomerization takes place when the butene–metal ligand is formed or when it is broken, i.e. which of the two reactions

$$PdBu_1 + Bu_1 \rightarrow PdBu_1Bu_2 \rightarrow PdBu_1 + Bu_2$$

or

$$PdBu_1 + Bu_1 \rightarrow PdBu_1Bu_1 \rightarrow PdBu_1 + Bu_2$$

is involved.

Organometallic catalysts of aluminium, the Alfin catalysts, are effective in polymerizing olefins, the mechanism of growth involving a displacement reaction.

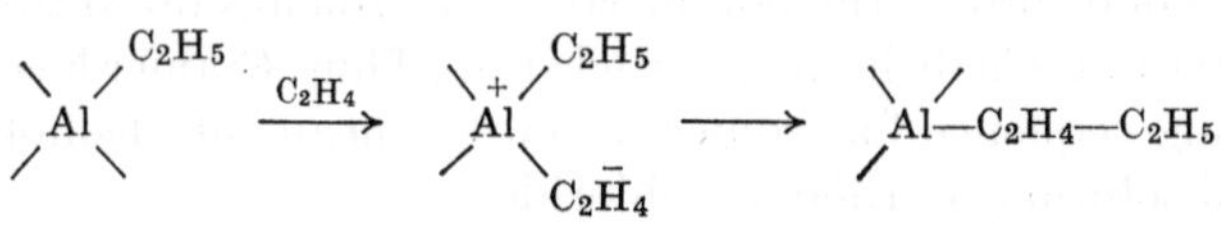

More effective than the Alfin are those of Ziegler and of Natta which consist of coordinated organometallic complexes of aluminium with a halide of a transition metal ($TiCl_3$, VCl_3, $CrCl_3$, $CoCl_2$). It is believed that the octahedrally coordinated Ti^{III} and the aluminium trialkyl which form the solid catalyst have two Ti—Cl ligands protruding from the surface of the solid; one halide ligand is attached to another Ti^{III} and is inactive, and the other can be replaced by an alkyl radical derived from aluminium trialkyl. In addition, there is one vacant coordination site. The olefin is inserted into this site and the insertion reaction takes place, permitting the reformation of the vacant site.

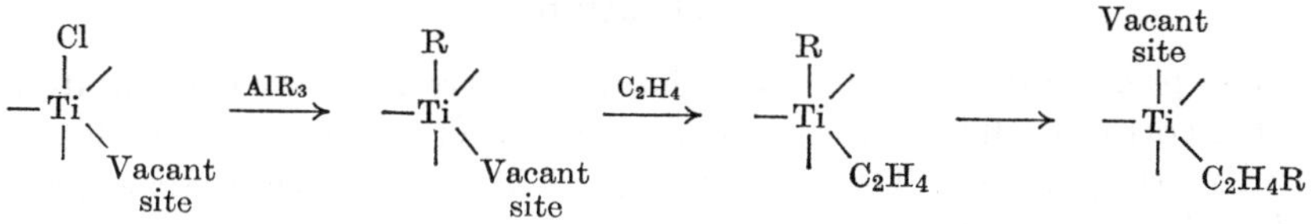

Chapter 8

Organic Syntheses and Decompositions

Cyclization

The passage of paraffins with chains of more than six carbon atoms or the corresponding olefins over several catalytic metals and oxides can lead to ring closure. The variation in yield of aromatic hydrocarbons when different initial paraffins are used can be predicted on the assumption that a mono-olefinic intermediary is held by two-point contact and that the double bond can migrate rapidly to all positions in the chain. Thus with n-hexane the following three configurations are possible for the adsorbed species:

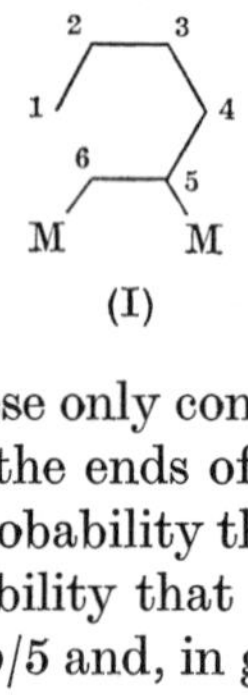

(I)

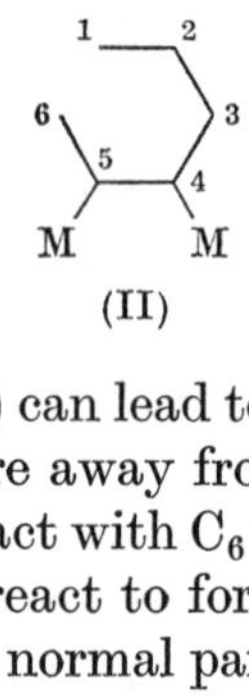

(II)

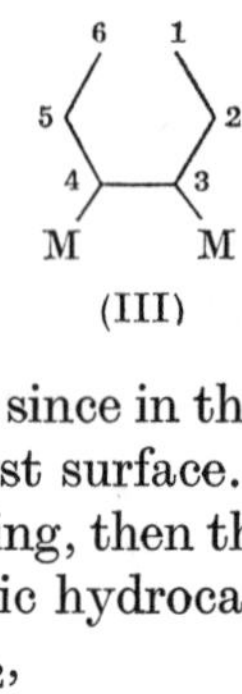

(III)

Of these only configuration (I) can lead to ring closure since in the other cases the ends of the chain are away from the catalyst surface. If p is the probability that C_1 will react with C_6 to close the ring, then the total probability that hexane will react to form an aromatic hydrocarbon is $P = 2p/5$ and, in general, for a normal paraffin C_nH_{2n+2},

$$P = \frac{2(n-5)p}{(n-1)}$$

if interaction with a tertiary carbon atom, e.g.

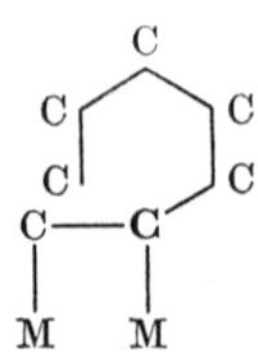

cannot take place.

For paraffins other than normal, e.g. with 2,5-dimethylhexane

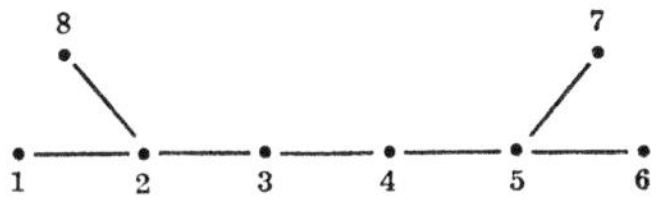

There are $(n-1)=7$ ways of choosing adjacent carbon atoms. If C_1 and C_2 are attached to the catalyst, then C_1 can react with C_8 or C_6. Thus, the probability for the pair is $2p$. There are four such pairs C_1C_2, C_2C_7, C_5C_6, C_5C_8, and no other cases lead to aromatization, hence $P=8p/7=1{\cdot}14p$.

The data in Table XLIV indicate how closely the experimental conversions agree with those calculated on this simple hypothesis.

TABLE XLIV

Paraffin	P/p	Conversion to aromatic (%) Obs.	Calc.
n-Hexane	0·40	20	23
n-Heptane	0·66	36	35
2-Methylhexane	0·66	31	35
n-Octane	0·84	46	42
3-Methylheptane	0·84	35	42
2,5-Dimethylhexane	1·14	52	52
n-Nonane	1·00	58	48

During ring closure isomerization can take place. It is found that the reaction cyclohexene → methyl cyclopentane can take place with great ease, and it appears that five–six ring isomerization is a very general reaction.

Although cyclohexane is adsorbed with the ring parallel to the surface, on dehydrogenation it passes through the rib form, i.e. a reorientation takes place. With methylcyclohexane this is much more difficult, but for isomerization which takes place on alumina at 40°C the rib configuration is necessary.

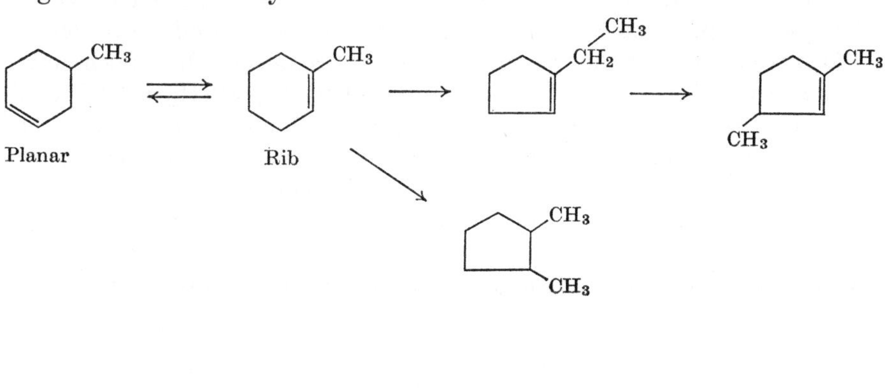

Hydrocarbon-reforming reactions include not only a number of reactions that can be ascribed to carbonium-ion formation, such as chain break-down and skeletal re-arrangement, but also dehydrogenation and ring closure resulting in the formation of aromatics. The hydrocarbon admixed with hydrogen is passed at high pressures and temperatures over the reforming catalysts. As a typical example we may cite the results of passing a n-heptane–hydrogen mixture (1:5·3 molar ratio) at 200 lb in^{-2} and a temperature of 500°C over an alumina supported platinum catalyst: The following fractions (mole %) were obtained: hydrocracked, 54·4; isomerized, 14·2; dehydrocyclized, 26·8; unchanged, 4·6.

Although platinum deposited on γ-alumina or produced by reduction of a platinum zeolite derived by base exchange in a calcium zeolite are effective catalysts, their relative activities for aromatization on the one hand and cracking and isomerization on the other, vary very considerably with the method of preparation. This variation can be ascribed mainly to the degree of dispersion of the platinum. Whereas it is believed that atomically dispersed platinum is formed by the base-exchange method, it is known that migration and sintering of deposited platinum can occur readily at the temperatures of catalytic activity. The "surface" of platinum has been evaluated by adsorption methods which include hydrogen, carbon monoxide and benzene as adsorbates. The rate of surface increase by sintering is found to be that of a second order equation or

$$\frac{1}{A} = \frac{1}{A_0} + kt$$

In an active reforming catalyst the crystallites are considered to be cubes of side some 15 Å and 200 $m^2\ g^{-1}$ Pt, exposing five surfaces of the cubes to the reacting gases. Sintering at 783°C will reduce the surface area to some 48 $m^2\ g^{-1}$ Pt in 10 h and to 5 $m^2\ g^{-1}$ Pt in 75 h and thus the catalytic efficiency for dehydrocyclization, which is found to be proportional to the surface of the platinum, will suffer a marked decrease. The second order rates of sintering have not been examined in detail but it would appear that the rate constants are affected by the circumambient gas: it is possible that a complex of the type PtH_2 or $Pt(CH_3)_2$ is more mobile than the chemisorbed atom of platinum. The effects of CO on the mobility of nickel and of hydrogen on copper indicate the importance of surface compounds in weakening the metal–metal bond. There is some experimental evidence for the existence of an optimum in the crystallite size for maximum catalytic activity. In support of such a view it may be noted that more complex planes can only grow as the crystallite grows from an elementary cube and, whereas ortho–para reaction can take place over the whole surface, reactions involving two- or three-point contact are restricted to certain planes with favourable atomic dispositions.

Aromatic Addition and Re-arrangements: The Raecke Reaction

Amongst aromatic rearrangements the carboxyl migration in the Raecke reaction is one of the most interesting. The conversion of *o*-phthalic acid to terephthalic acid proceeds smoothly with high yields (93–94%) on heating the potassium salt in an atmosphere of carbon dioxide at 100 atm up to some 450°C. The addition of a small quantity of cadmium fluoride increases the yield slightly. The sodium salt gives smaller yields and the lithium salt is least effective. Carboxyl migration is likewise involved in disproportionation. Dipotassium terephthalate and benzene are formed by disproportionation of potassium benzoate utilizing cadmium fluoride as catalyst at 400°C under pressure. Numerous variations of these migratory and disproportionating reactions under high pressures of carbon dioxide have been carried out. Cadmium and zinc salts are found to be the most effective catalysts and the anionic efficiencies were found to be $I^- > Br^- > Cl^- > CO_3^{2-}$.

The reaction path probably involves a primary decarboxylation followed by a recarboxylation, since it has been claimed that both benzene and toluene can react directly with carbon dioxide at high pressure utilizing aluminum chloride as a catalyst to yield the respective benzoic and *p*-toluic acids. The further conversion of benzoates to terephthalic acid can be carried out in an analogous manner.

Ammonoxidation

The first example of the process of ammonoxidation was in the successful conversion of propane into acrylonitrile $CH_3CH_2CH_3 \rightarrow CH_2{=}CH{-}CN$ by means of a vanadium oxide catalyst utilizing a gaseous mixture of air and ammonia. In a similar manner it has been found possible to convert toluene to benzonitrile with a high efficiency over a vanadium oxide catalyst supported on alumina. In all probability the reaction proceeds through the intermediary of an aldehyde

$$CH_3CH_2CH_3 \longrightarrow CH_2{=}CHCH_3 \longrightarrow CH_2{=}CH{-}CHO \longrightarrow CH_2{=}CHCN$$

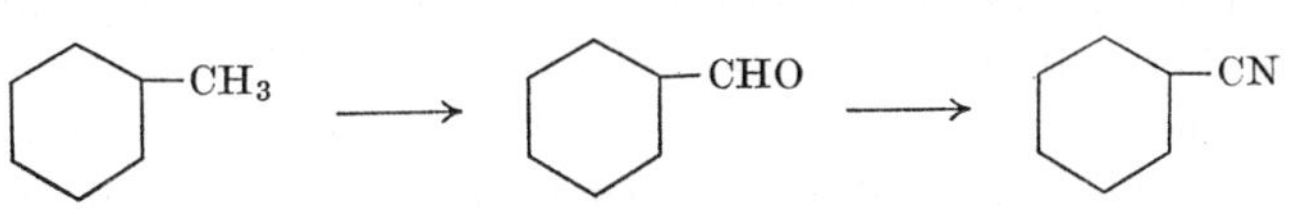

On the ammonoxidation of *o*-xylene over a mixture of molybdenum and vanadium oxides at about 380°C phthalimide can be produced with a relatively high yield.

Acrylic Acid

The textile demand for acrylates has led to the development of a number of catalytic processes for their synthesis. As raw materials acetylene, ethylene oxide, acetic acid, formaldehyde, and propylene are utilized. The original Reppe process involved the direct synthesis from acetylene, carbon monoxide and water over a nickel catalyst. The nickel catalyst can be replaced by nickel carbonyl, and esters can be formed directly by replacement of the water by an alcohol.

When ethylene oxide is used as the source it is converted into ethylene cyanohydrin by reaction with hydrocyanic acid in the presence of a base. The cyanohydrin is subsequently converted into acrylic acid by hydrolysis. The conversion of acetic acid into ketene can be effected by dehydration and this reacts directly with formaldehyde to yield β-propiolactone which, with acid catalysts such as phosphoric acid, is converted into acrylic acid.

We have already noted the conversion of propane into acrylonitrile by ammonoxidation; propylene can be converted into acrylic acid by direct oxidation. Two methods of effecting this oxidation have been proposed, either in the liquid phase at temperatures below 100°C (in which alkaline suspensions of silver and silver oxide are used as catalysts) or in the vapour phase at ca 500°C (with catalysts such as cobalt molybdate). The oxidation proceeds through the intermediary formation of acrolin.

The Decomposition of Formic Acid

Formic acid can undergo two methods of catalytic decomposition to yield either carbon dioxide or the monoxide according to the equations

$$HCOOH \rightarrow CO_2 + H_2$$
$$HCOOH \rightarrow CO + H_2O$$

Of these, decomposition to the dioxide is more common, and proceeds almost exclusively so on many metals; dehydration can be effected on oxides.

On metals formic acid undergoes chemisorption to yield metallic formates. The salts of silver and gold are less stable than that of nickel which has a high heat of formation. The catalytic reaction involved, being essentially that of the thermal decomposition of the surface formate, is in consequence more complex on the former. On silver the energies of activation have been found to vary from ~ 16 kcal mole^{-1} to 25 kcal mole^{-1}, depending on the preparation on the silver, and it is believed that the value of 25 kcal mole^{-1} is associated with the 100 facet. On gold a value of 14·3 kcal mole^{-1} and on copper 28·1 kcal mole^{-1} have been reported. Part of the difficulty associated with the low temperature decomposition on gold and silver may be ascribed to the fact that if the reaction mechanism is written

$$HCOOH+M \rightarrow MHCOO^- + H_a$$
$$MHCOO^- \rightarrow CO_{2_g} + MH \rightarrow M + H_a$$
$$2H_a \rightarrow H_2$$

the surface must contain adsorbed hydrogen atoms which have to be eliminated by the reaction $2H_a \rightarrow H_2$. In the steady state the reaction velocity is accordingly

$$\omega = K\theta_{HCOO^-} = K^1\theta_H^2$$

where $\theta_{HCOO^-} + \theta_H = 1$. Unless the recombination of the hydrogen atoms is fast compared to the rate of decomposition of the formate, a change in temperature will alter the surface composition and thus the apparent energy of activation.

If formic acid is admitted to a nickel surface at 100°C the vapour is at first adsorbed and subsequently there is a liberation of hydrogen and the formation of a formate. It is only at the end of the second stage that carbon dioxide appears in the effluent gas. Eventually, carbon dioxide and hydrogen are liberated in the equivalent amounts. The primary formation of a metal formate can be confirmed in many ways, e.g. by infrared analysis. Although the nickel formate is sufficiently stable at low temperatures as to ensure a zero order reaction, at higher temperatures it becomes first order.

There does not appear to be any interaction between surface formate groups, e.g.

$$2MHCOO \rightarrow 2M + H_2 + 2CO_2$$

for if formic acid is admixed with deuterium and decomposed on a gold catalyst which does not catalyse the hydrogen–deuterium reaction no HD is formed but equilibrated HD is obtained when mixtures of HCOOH and DCOOD are decomposed. Schwab and Watson have recently examined the decomposition of the acid and its three deuterated forms on a number of metals and have noted that the differences in energies of activation of the deuterated forms are greatly dependent on the entropies of activation for the formation of the transition complexes as will be noted in Table XLV.

TABLE XLV

Entropies of activation
(cal mole^{-1} deg^{-1})

Acid	Au	Hg	Cu
HCOOH	−61·1	−53·2	−35·4
HCOOD	−58·6	−50·1	−33·6
DCOOH	−55·1	−49·2	−33·1
DCOOD	−53·7	−46·3	−28·5

On oxides and semiconductors the decomposition likewise proceeds through the formation of formates which can be formulated in the following way.

$$HCOOH + MO \rightarrow MHCOO + OH \rightarrow H_2 + CO_2$$

The activation energy for the surface salt formation evidently involves the electron-accepting power of the semiconductor. Such decompositions should therefore proceed more readily on a p-type rather than on an n-type oxide.

The oxide NiO is catalytically active and has been regarded as functioning as a p-type catalyst. The number of unpaired electrons in NiO can be reduced by means of a support such as Al_2O_3 and further reduced by addition of divalent cations and increased with quadrivalent ones. The changes in activation energy accompanying changes in catalyst composition are for NiO, 26·5 kcal mole^{-1}; for NiO on Al_2O_3, 20·5 kcal mole^{-1} and for NiO on $Al_2O_3 + 5\%\ TiO_2$, 24 kcal mole^{-1}.

In view of the ease with which the oxide can at least superficially be reduced, these changes in activation energy may well bear re-examination. On oxides such as MgO, Al_2O_3 and SiO_2 the dehydrogenating reaction is accompanied by one of dehydration to yield carbon monoxide. The relative rates of the two reactions are usually described as a selectivity ratio.

$$S = \frac{CO_2 \text{ rate}}{CO \text{ rate} + CO_2 \text{ rate}}$$

For magnesia Sabatier found $S = 50\%$ but values have been found for different samples from $S = 15$ to $S = 90\%$. High S values are favoured by high temperatures.

The mechanism of the dehydration reaction is not so well established. According to some investigators the S values for bulk formate decompositions derived from various oxides, are the same as for the catalytic decomposition of the acid on the same oxides, but others find that this is not the case: thus Stone found $S = 60\,\%$ for the decomposition of zinc formate but carbon dioxide only was found in the catalytic decomposition of the acid on zinc oxide. On acidic oxides the dehydration reaction is regarded as involving a proton addition to an adsorbed acid molecule.

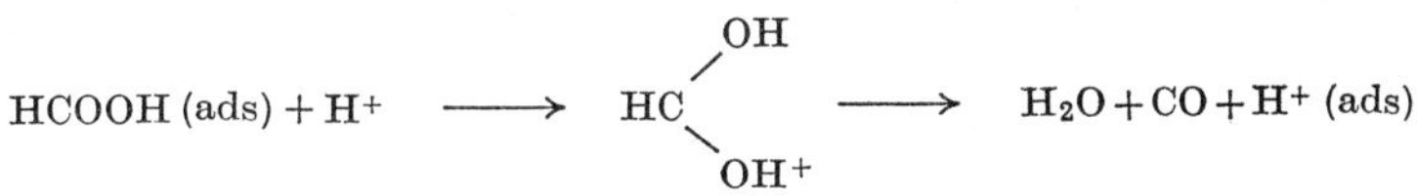

and, in an analogous manner, we may regard the dehydration on other oxides as involving the metallic ion

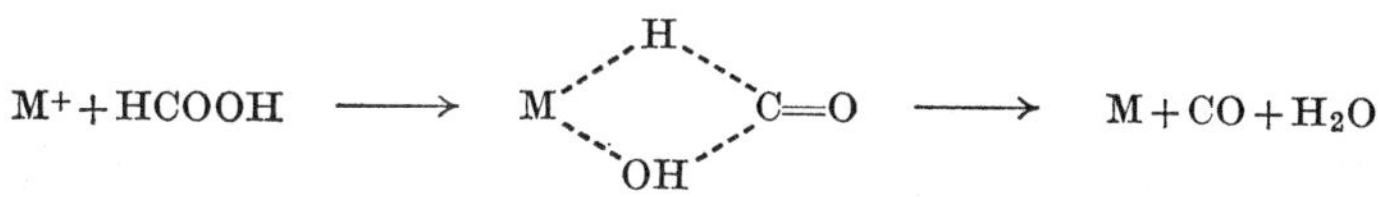

Decomposition of Alcohols

The aliphatic alcohols can undergo two methods of catalytic decomposition on metallic oxides. On some oxides dehydration results in the formation of an olefin, e.g.

$$CH_3CH_2CH_2OH \rightarrow CH_3CH{=}CH_2 + H_2O$$

On other oxides the product is an aldehyde, e.g.

$$CH_3CH_2CH_2OH \rightarrow CH_3CH_2CHO + H_2$$

Oxides such as thoria or alumina yield olefins and others such as cadmium or magnesium oxides produce aldehydes. Again oxides such as beryllium or zirconium oxide are found to give almost equal quantities of the two products.

It is found that on the dehydration of ethyl alcohol on alumina as the temperature is lowered there is an increasing quantity of diethyl ether found in the products. Diethyl ether itself is readily dehydrated in turn to ethylene. This can follow a bimolecular process or one of disproportionation.

$$2C_2H_5OC_2H_5 \rightarrow 2C_2H_4 + H_2O$$

$$C_2H_5OC_2H_5 \rightarrow C_2H_4 + C_2H_5OH$$

It is believed that olefin production proceeds through the formation of an oxonium carbonium ion, the necessary proton being derived from a Brønsted site on the catalytic oxide

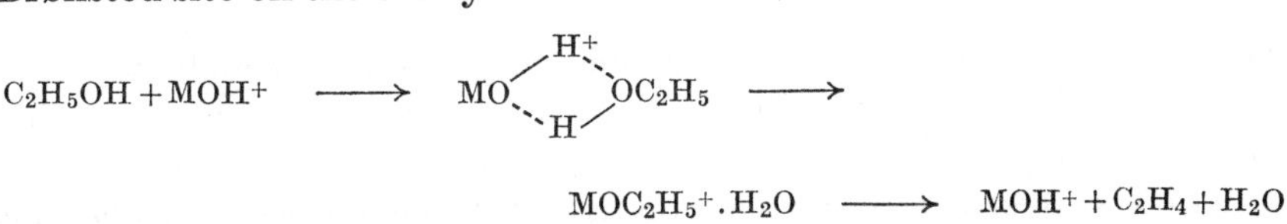

On this view the intermediary ether can be formed by reaction of gaseous alcohol with the oxonium ion.

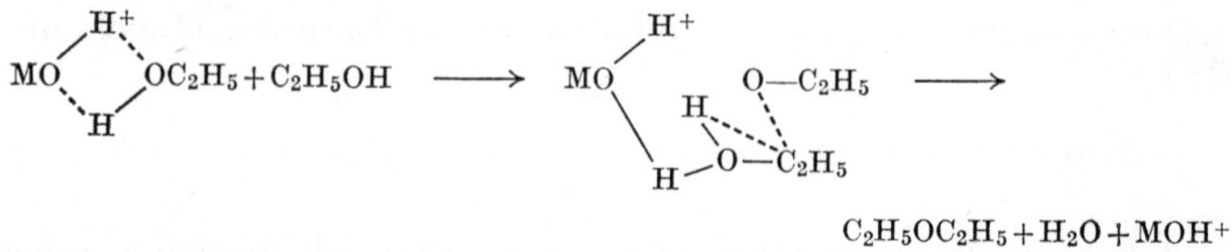

The formation of ether can be by-passed at higher temperatures by the direct abstraction of water

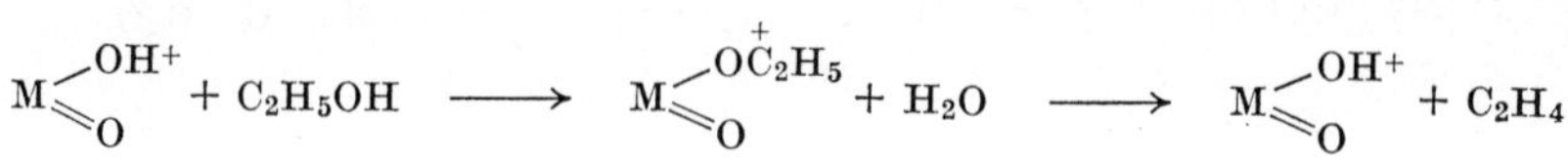

a reaction analogous to Williamson's scheme for etherification. That a carbonium ion is an intermediary in dehydration process receives support from the fact that isomeric olefin mixtures are found in the products from the higher alcohols, e.g.

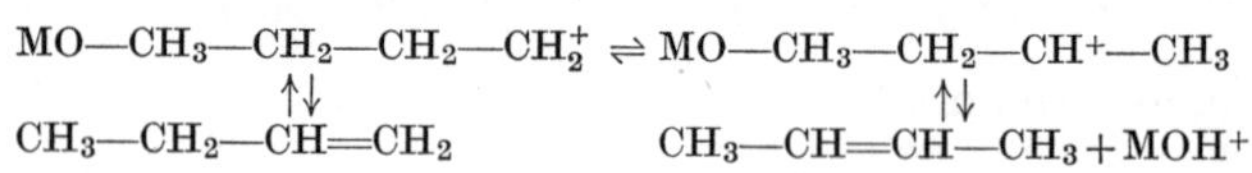

In addition, those oxides that adsorb proton acceptors such as pyridine are more active for dehydration than catalysts for dehydrogenation such as magnesia which adsorb proton donors, e.g. phenol, more readily. Dehydrogenation on semiconducting oxides is favoured with those of the p-type.

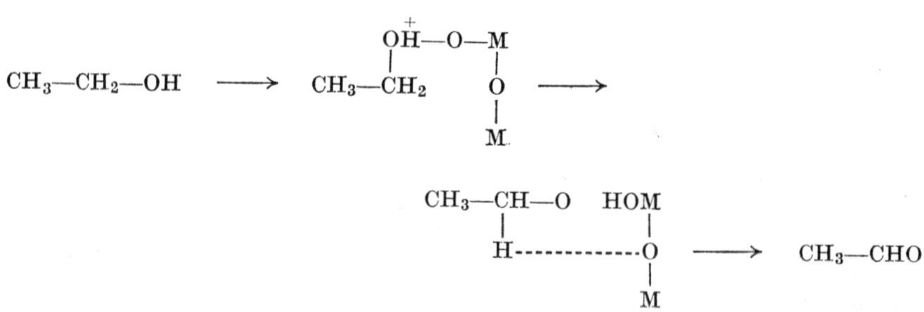

On the assumptions that both ethanol and water are strongly adsorbed by the catalyst in comparison with either ethylene or ether and that each obeys Langmuirian adsorption on one elementary site, the rate of reaction is governed by the reaction rate of the chemisorbed alcohol, i.e.

$$\omega = K\theta_{C_2H_5OH}$$
$$= \frac{K' p_{C_2H_5OH}}{1 + bp_{C_2H_5OH} + b' p_{H_2O}}$$

The reaction is thus strongly inhibited by water and, in its absence, the reaction would be of zero order in respect to alcohol over a wide range of pressures.

There is great variation in the values given for the apparent energies of activation for the dehydration process, varying from 12 to some 40 kcal mole^{-1}; such is also the case for dehydrogenation reactions although these are on the whole somewhat lower than the former. Thus, on zinc oxide $E_{\text{dehydrogenation}} = 17–25$ kcal mole^{-1} and $E_{\text{dehydration}} = 37–56$ kcal mole^{-1}.

A potentially important synthesis of butadiene from ethyl alcohol was discovered by Ostromislenski, who showed that passage of ethyl alcohol over a zinc oxide–alumina catalyst at 400°C resulted in the formation of buta-1,3-diene. The mechanism involved consists in the formation of acetaldehyde which undergoes condensation to aldol; the aldol at these temperatures undergoes dehydration to buten-2-al which in turn by processes involving dehydrogenation to 1-hydroxybut-2-ene and dehydration results in the formation of $CH_2{=}CH{-}CH{=}CH_2$.

The above mentioned processes of dehydration and dehydrogenation are all reversible and the hydration of ethylene to yield derivatives such as ethyl alcohol, acetaldehyde or acetic acid as well as propan-2-ol, butanol and acetone is the basis of technically important operations. In general, acid phosphates as catalysts have been preferred to the oxides; indeed, it has been claimed that cadmium metaphosphate will equilibrate a gaseous mixture of ethanol, ethylene and steam at a temperature as low as 175°C. The equilibrium constant $K_p = [C_2H_5OH]/[C_2H_4][H_2O]$ is given by the following relationship

$$\log_{10} K_p = \frac{2000}{T} - 6{\cdot}14$$

The Water Gas Shift Reaction

Apart from the electrolytic production of hydrogen, either hydrocarbons or water gas may be employed for its manufacture by catalytic processes.

From hydrocarbons, hydrogen may be produced by catalytic interaction with steam, based upon the two reactions

$$C_nH_{2n+2} + 2nH_2O \rightarrow nCO_2 + (3n+1)H_2$$
$$C_nH_{2n+2} + nH_2O \rightarrow nCO + (2n+1)H_2$$

and the further utilization of the carbon monoxide by the water gas shift reaction

$$CO + H_2O \rightleftharpoons CO_2 + H_2$$

The customary steam–iron process in which a water gas–steam cycle is utilized to bring about the alternate oxidation and reduction of spathose, a chalybite oxide of iron, is now replaced by a catalytic process involving the water gas shift. In either case, carbon dioxide removal can be effected by pressure water scrubbing and residual carbon monoxide by washing with cuprous ammonia solution, by catalytic hydrogenation to comparatively inert methane or by selective low-temperature combustion to carbon dioxide. The oxidation and reduction of the chalybite involves the following approximate heat changes:

$$\begin{aligned} Fe_3O_4 + H_2 &\rightleftharpoons 3FeO + H_2O & \Delta H &= 27{,}300 \text{ B.t.u. lb mole}^{-1} \\ Fe_3O_4 + CO &\rightleftharpoons 3FeO + CO_2 & \Delta H &= 13{,}200 \text{ B.t.u. lb mole}^{-1} \\ FeO + H_2 &\rightleftharpoons Fe + H_2O & \Delta H &= 6500 \text{ B.t.u. lb mole}^{-1} \\ FeO + CO &\rightleftharpoons Fe + CO_2 & \Delta H &= -7700 \text{ B.t.u. lb mole}^{-1} \end{aligned}$$

The reaction $CO + H_2O \rightarrow CO_2 + H_2$ is consequently exothermic with $\Delta H = 14{,}000$ B.t.u. lb mole^{-1} at 900°C or 10,090 cal mole^{-1} at 290°C, and thus it is important to find catalysts that are active at low temperatures, so as to ensure the maximum utilization of the carbon monoxide. It has been claimed that a catalyst of composition 94% Co, 5% Al and 1% K will equilibriate a gas mixture at about 300°C. Taking the conventional chemical constants as $i_{CO} = -0{\cdot}05$, $i/_{H_2O} = -1{\cdot}90$, $i/_{CO_2} = -0{\cdot}90$, $i/_{H_2} = +3{\cdot}68$, then

$$\log K_c = -\frac{9802}{4{\cdot}576T} + \frac{0{\cdot}99}{1{\cdot}987}\log T + \frac{1}{4{\cdot}576}\int_0^T \frac{\partial T}{T^2}\int_0^T \Sigma\, C_v\, \mathrm{d}T + 0{\cdot}83$$

and the approximate values of K_p, the equilibrium constant,

$$K_p = \frac{p_{H_2O}\, p_{CO}}{p_{CO_2}\, p_{H_2}}$$

are as follows: 0·0534 at 355°, 0·140 at 454°, 0·323 at 573°, 0·460 at 633°, 0·825 at 771°, 1·147 at 860°, 1·671 at 986°C.

Whereas many oxides and metals can bring about the water gas shift at high temperatures, more active ones (usually with an iron oxide base to which chromia and in some cases traces of alkali are added) will effect equilibrium at temperatures as low as 450°C.

No clear cut experiments have been recorded as to the mechanism of the shift reaction. Although the oxide undergoes alternate oxidation and reduction the sequence of events is far from clear. If magnesium oxide is used as a catalyst, small quantities of magnesium formate can be detected in the catalyst after use, and this suggests that the formate ion may play the part of an intermediary. We could on this view write the sequence of events as follows.

$$MgO + H_2O \rightleftharpoons Mg\begin{matrix} OH \\ OH \end{matrix} \tag{26}$$

$$Mg\begin{matrix} OH \\ OH \end{matrix} + CO \rightleftharpoons Mg\begin{matrix} OH \\ COOH \end{matrix} \tag{27}$$

$$Mg\begin{matrix} OH \\ COOH \end{matrix} \longrightarrow CO_2 + Mg\begin{matrix} H \\ OH \end{matrix} \tag{28}$$

$$Mg\begin{matrix} H \\ OH \end{matrix} \longrightarrow MgO + H_2 \tag{29}$$

It is possible that the basic formate formed by a typical carbon monoxide insertion reaction (27) undergoes re-arrangement to a carbonate before the liberation of the carbon dioxide and that hydrogen is first eliminated, e.g.

$$Mg\begin{matrix} OH \\ COOH \end{matrix} \longrightarrow MgCO_3 + H_2 \tag{28a}$$

$$MgCO_3 \longrightarrow MgO + CO_2 \tag{29a}$$

Some possible reaction mechanisms can be dismissed by utilizing isotopic reactants. Kaneko has made use of Horiuti's method for isolating the rate-determining step by determination of its stoichiometric number. The basis of the method resides on the premise that isotopic transfer takes place only through the rate determining step of the overall reaction. The free energy of the water gas reaction is given by:

$$\Delta G = RT \log\left(\frac{p_{CO_2} p_{H_2}}{p_{CO}\, p_{H_2O}}\right) + RT \log K_p \tag{30}$$

and the rate of the reaction in a mixture of carbon monoxide and dioxide, D_2O and hydrogen would be given by the difference in the velocities of the forward and back reactions

$$\omega = \vec{\omega} - \overleftarrow{\omega} = \frac{dn_{H_2O}}{dt} \tag{31}$$

By following the change in deuterium content of the water as the reaction approaches equilibrium, the velocity of transfer of deuterium to hydrogen can be determined, i.e.

$$\frac{dn_{H_2O}}{dt} a_{H_2O} = a_{H_2O}\vec{\omega} - a_{H_2}\overleftarrow{\omega} \tag{32}$$

Where a_{H_2O} and a_{H_2} are the atomic fractions of deuterium in the water and in the hydrogen. From equations (31) and (32) we obtain:

$$\frac{\vec{\omega}}{\overleftarrow{\omega}} = 1 = \left(1 - \frac{a_{H_2}}{a_{H_2O}}\right) \frac{d \log n_{H_2O}}{d \log a_{H_2O}} \tag{33}$$

The stoichiometric number $n(r)$ is given by the equation:

$$n(r) = \frac{\Delta G}{RT} \log \frac{\vec{\omega}}{\overleftarrow{\omega}}$$

which can be evaluated by means of (30) and (33).

Experiments with both deuterium and ^{14}C gave results in conformity with a value of $n(r)$ of unity in both cases. Consequently, the rate-determining step must involve both carbon monoxide and water vapour, i.e. it must follow some pattern involving one of the reactions:

$$\left.\begin{array}{r} CO(ads) + H_2O \\ CO + H_2O(ads) \\ CO(ads) + H_2O(ads) \end{array}\right\} \rightarrow CO_2 + H_2$$

With the scheme proposed for reactions with magnesium oxide as catalyst the rate-determining reaction would be either (27), (28), or (28a).

Synthesis of Alcohols

Sabatier and Senderens were the first to investigate the possibilities inherent in the catalytic conversion of water gas into methyl alcohol according to the equilibrium

$$CO + 2H_2 \rightleftharpoons CH_3OH$$

The equilibrium constant of the reaction

$$\log_{10} K_p = \log \frac{p^2_{H_2} p_{CO}}{p_{CH_3OH}}$$

has the following approximate values in logarithms: 1·24 at 400°, 1·85 at 500°, 2·94 at 600°, 4·72 at 700°, 5·90 at 800°, 7·80 at 1000°K. At 600°K the equilibrium constant is $\sim 10^3$, and the effect of pressure on the alcohol content at equilibrium from an equimolecular mixture of carbon monoxide and hydrogen is shown in Table XLVI.

TABLE XLVI

Total pressure (atm)	p_{H_2} (atm)	p_{CO} (atm)	p_{CH_3OH} (atm)
1	0·5	0·5	$9{\cdot}5 \times 10^{-2}$
10·1	5	5	0·125
22·5	50	50	125

Thus, effective catalysts must operate at relatively high pressures and low temperatures. It was found that a variety of catalysts was capable of synthesizing small quantities of methyl alcohol, but higher alcohols and hydrocarbons could also be produced. In consequence the investigations proceeded along three different paths, firstly with the object of obtaining methyl and other alcohols, secondly to produce oxy derivatives of hydrocarbons, the so-called oxo process, and finally the conversion of water gas into liquid hydrocarbon fuels, the Fischer Tropsch process. Although alcohols can be produced in all these reactions the catalysts developed for each process which has its own objectives are very different.

Methyl alcohol

The most effective catalyst for the synthesis of methyl alcohol is pure zinc oxide. In the synthesis traces of higher alcohols are almost invariably present and these can be increased to a remarkable extent by the incorporation of alkali such as K_2O or rubidium or caesium oxide; chromium oxide Cr_2O_3 is likewise an effective catalyst, and $ZnO.Cr_2O_3$ and zinc chromite mixtures are generally employed. The addition of basic oxides serves to promote the production of higher alcohols. For methyl alcohol synthesis the temperature must be maintained below 400°C, above which there is a sharp increase in production of the higher alcohols. Both zinc oxide and chromium oxide are p-type semiconductors; the latter is converted above 400°C into an n-type.

The simplest mechanism to account for the synthesis is reaction of hydrogen with chemisorbed carbon monoxide to form a transition complex.

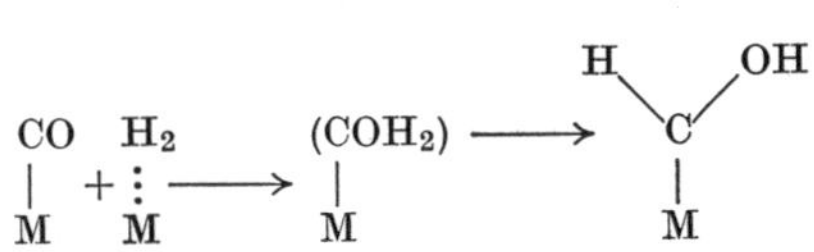

If the C—M bond is still a strong one after reaction, water will be removed and the residue will be a "surface carbide" which plays an important part in methane formation as well as in the Fischer Tropsch process. If, however, the bond is a weak one, formaldehyde may result or methyl alcohol may be formed by interaction

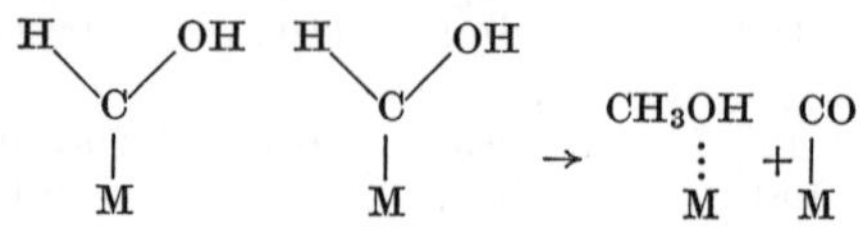

reforming a molecule of carbon monoxide.

Higher alcohols

The incorporation of alkali in the zinc oxide results in an increase in the concentration of higher alcohols. The alkaline zinc oxide catalysts have but a short life and stability. These can be improved by promotion, effected by the incorporation of other oxides. In addition to chromium oxides, ternary systems such as zinc, copper, chromium and manganese–zinc–chromium oxides have been utilized. Further promotion by means of potassium oxide yields catalysts of remarkable activity. These catalysts are usually prepared from the precipitated hydroxides or carbonates.

Several schemes have been suggested for the production of the higher alcohols. The simplest proposal is the suggestion that the methyl alcohol is stabilized on the surface as potassium methylate CH_3OK(ads) which then undergoes the carbon monoxide insertion reaction

$$CH_3OK(ads) + CO \rightarrow CH_3COOK(ads)$$

the acetate being subsequently reduced to ethanol

$$CH_3COOK(ads) + H_2 \rightarrow CH_3CH_2OK(ads)$$
$$CH_3CH_2OK(ads) + H_2O \rightarrow CH_3CH_2OH + KOH(ads)$$

The potassium acetate can likewise undergo thermal decomposition to acetone which on reduction yields isopropyl alcohol.

The oxo reaction

The higher alcohols can also be produced from olefins by hydroformylation which is analogous to the hydrogenation process with the insertion of a CO group in the molecule

$$CH_2{=}CH_2 + H_2 \rightarrow CH_3{-}CH_3$$
$$CH_2{=}CH_2 + CO + H_2 \rightarrow CH_3CH_2CHO$$

The olefins generally employed are propylene and its polymers as well as butylenes and the interpolymers. The heat of olefin hydroformylation is some $\Delta H = -35$ kcal mole^{-1}, and the free energies and equilibrium constants are given in the Table XLVII.

TABLE XLVII

Temp. (°K)	$-\Delta G$ (kcal mole^{-1})	K_p
298	14·46	$4{\cdot}03 \times 10^{10}$
373	8·93	$1{\cdot}11 \times 10^{5}$
423	5·28	$4{\cdot}99 \times 10^{2}$
473	1·39	4·39

$$K_p = \frac{[CH_3CH_2CHO]}{[C_2H_4][CO][H_2]}.$$

Somewhat surprisingly, in spite of the greater change in free energy —$\Delta G = 22{\cdot}6$ kcal mole^{-1}, hydrogenation of the double bond is practically eliminated in the presence of carbon monoxide although with conjugated diolefins saturated hydrocarbons as well as aldehydes and alcohols are found in the products. The oxo reaction is usually carried out in a liquid phase with pressures from 70 up to 250 atm and at temperatures below 250°C. Whilst the earlier catalysts were based upon the catalysts developed for the Fischer Tropsch hydrocarbon synthesis, namely those containing cobalt and thoria on various support materials such as magnesia and Kieselguhr, it was soon discovered that hydrocarbon synthesis could be reduced and alcohol production raised if cobalt alone was employed, and the standard method of operation employs a soluble cobalt carbonyl as a catalyst. Thus an organometallic catalyst similar to the organometallic ruthenium catalysts used for hydrogenation now replaces the original heterogeneous catalysts. Hydroformylation can be effected by the use of dicobalt octacarbonyl which in the presence of hydrogen dissociates to form cobalt hydrotetracarbonyl.

$$H_2 + Co_2(CO)_8 \rightleftharpoons 2HCo(CO)_4 \qquad (34)$$

The mechanism of the catalytic intrusion of the carbonyl group is most readily interpreted on the basis that the formation of a tricarbonyl by dissociation is necessary

$$HCo(CO)_4 \rightleftharpoons HCo(CO)_3 + CO \qquad (35)$$

We must note that tricarbonyl formation is cut down as the pressure of the carbon monoxide increases. The tricarbonyl can now make a π-complex with the olefin.

$$HCo(CO)_3 + C_2H_4 \rightleftharpoons HCo(CO)_3.C_2H_4 \quad k_1 \qquad (36)$$

which can then re-arrange to give an alkyl cobalt tricarbonyl.

$$HCo(CO)_3.C_2H_4 \rightleftharpoons Co(CO)_3.C_2H_5 \qquad (37)$$

which in its turn can be converted into an unstable tetracarbonyl by means of carbon monoxide.

$$Co(CO)_3.C_2H_5+CO \rightleftharpoons Co(CO)_3(CO).C_2H_5 \quad k_2 \qquad (38)$$

This in turn reacts with hydrogen to give propionaldehyde and the tricarbonyl, thus completing the process of hydroformylation.

$$Co(CO)_3(CO).C_2H_5+H_2 \rightarrow Co(CO)_3(H_2)(CO)C_2H_5 \rightarrow HCo(CO)_3+CH_3CH_2CHO \quad k_3 \qquad (39)$$

The whole process thus involves the progressive addition of ethylene or other olefin, then carbon monoxide and finally hydrogen to the cobalt carbonyl of which the active species is the tricarbonyl. On this hypothesis of an active cobalt hydrotricarbonyl as catalyst the observed isomerization of olefins by reaction (36) is readily understood.

On the assumption that reactions (34), (35) and (37) are reversible and always in equilibrium, from reactions (36), (38) and (39) we obtain the rate equation

$$\omega = \frac{k_1 k_3 p_{H_2}}{k_2 p_{CO} + k_3 p_{H_2}} [HCo(CO)_3][C_2H_4]$$

which fits the experimental data on the velocities of reaction obtained with different $CO:H_2$ ratios at elevated pressures. The apparent energy of activation $E = 34$ kcal mole^{-1} at 100°C falling to 25 kcal mole^{-1} at more elevated temperatures. At low pressures $\omega \propto [C_2H_4]p_{H_2}$ up to 100 atm. At fixed p_{H_2} the velocity first rises and then falls with increasing p_{CO} as might be anticipated from (38) and (35). Homogeneous catalysts are usually employed in the insertion of a CO group into alcohols, acetylene derivatives and the like. These are embraced by the general term of carboxylation. In this way formates can be produced from alcohols and carbon monoxide utilizing the alcoholate ion as catalyst, e.g. with methyl alcohol and sodium methoxide we have:

$$CH_3O^- + CO \rightleftharpoons (COOCH_3)^-$$
$$COOCH_3^- + CH_3OH \rightleftharpoons HCOOCH_3 + CH_3O^-$$

The carbon monoxide in this reaction must be regarded as an electron acceptor or double bonded with an empty orbital on the carbon atom. The most active catalytic alcoholates are those in which the electron density at the oxygen atom is raised by substitution of the hydrogens in the methyl group with electron donors. On the other hand, in acid solution alcohols are carboxylated by CO insertion in the presence of nickel or cuprous halides with the corresponding halide acid at temperatures of 250–300°C and pressures above 200 atm.

$$CH_3OH + CO \rightarrow CH_3COOH$$

One of the most important of these carboxylation reactions is, as we have noted, the production of acrylic acid from acetylene and carbon monoxide, where nickel carbonyl in the presence of a halide acid is used as catalyst. Thus ethyl acrylate can be prepared from carbon monoxide and acetylene with the stoichiometric required quantity of alcohol utilizing nickel carbonyl as catalyst at pressures of ~30 atm and 180°C according to the following reaction.

$$4C_2H_2 + Ni(CO)_4 + 4EtOH + 2HBr = 4CH_2CHCOOH + 4NiBr_2 + H_2$$

Alcohols can also be produced by the selective reduction of fatty acids. This reduction can proceed in stages

$$RCOOH \rightarrow RCH_2OH$$
$$RCOOH \rightarrow RCH_3$$
$$RCOOH \rightarrow RH + CH_4$$

With active catalysts and at relatively low temperatures a prepondence of alcohols is obtained. If similar pressures of hydrogen are employed, it is found that increasingly higher temperatures are required to effect hydrogenation at comparable rates with metal and metal alloy catalysts as the *d*-band vacancies are progressively diminished. Whilst active nickel catalysts yield a fair proportion of alcohols at low temperatures even lead will catalyse the removal of the carboxyl group as methane at high temperatures.

The Fischer Tropsch Reaction

Carbon monoxide–hydrogen mixtures in contact with metallic iron, cobalt, ruthenium, or nickel yield hydrocarbons and aliphatic alcohols. The production of aliphatic alcohols is greatest at low temperatures, i.e. 200–300°C. On elevation of the temperatures the alcohol content of the product falls, that of the hydrocarbons increases but their chain length shortens. The rate of synthesis increases with increase in pressure but the production of non-volatile hydrocarbons of very long chain lengths is also facilitated, necessitating frequent dewaxing of the catalyst.

The activity of these catalytic metals can be augmented by promoters especially alkaline ones such as MgO and kieselguhr; indeed, with iron, admixture with an alkali is essential. Potassium is an effective promoter with 2·0 % K_2O. The mechanism of synthesis has not been satisfactorily established.

It is improbable that chemisorption of hydrogen is involved or that the metal itself is the active catalyst, for when synthesis gas is passed over the metal the first product of reaction is methane and no higher hydrocarbon. Occasionally, several hours of methane production are noted before the synthesis of hydrocarbon "oils" commences. When the catalyst is in this state, viz. capable of forming methane only, the ortho–para hydrogen conversion can also take place, but this reaction also ceases when the synthesis commences. Although carbides of the metals are readily formed by synthesis gas, of the two possible mechanisms of carbide formation, namely

$$M + 2CO \rightarrow MC + CO_2 \quad \text{the Boudouard reaction} \tag{40}$$

$$M + CO \rightarrow MCO \xrightarrow{H_2} MC + H_2O \tag{41}$$

reaction (41) is favoured since the reduction of chemisorbed carbon monoxide by hydrogen to water is a much more rapid reaction than the "carbiding" of the metal by carbon monoxide alone.

It is, however, unlikely that normal carbides are active catalysts since neither Co_2C nor Ni_3C are active but the carbides of iron Fe_3C and Fe_2C are reactive.

The active catalyst may be a "surface" carbide, e.g. of face-centred cubic nickel and not the hexagonal bulk nickel carbide, since the spacing on the cubic nickel is appropriate for the adsorption of linked methylene radicals. Olefins can be incorporated in the growing chains by direct addition to the synthesis gas, and it has been established using isotopic carbon that alcohols, aldehydes and ketene when added to the gas are found to have acted as chain initiators, since the isotope is terminal. The following mechanism has been proposed for the chain initiation and chain growth:

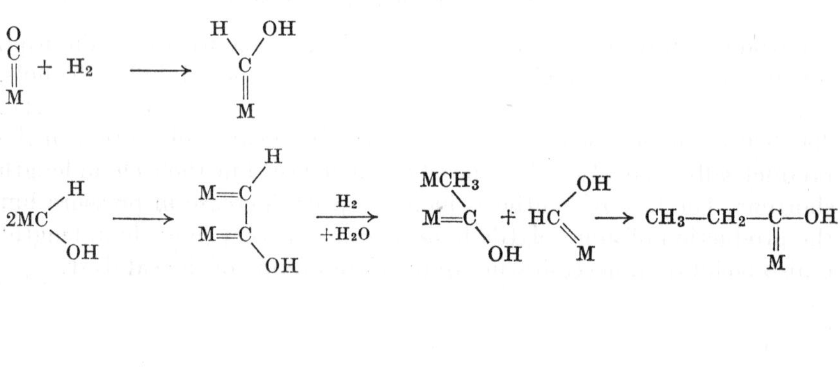

Although branching chains result from a further hydrogenation, viz.

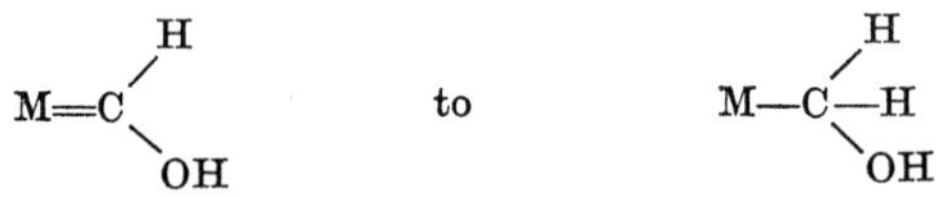

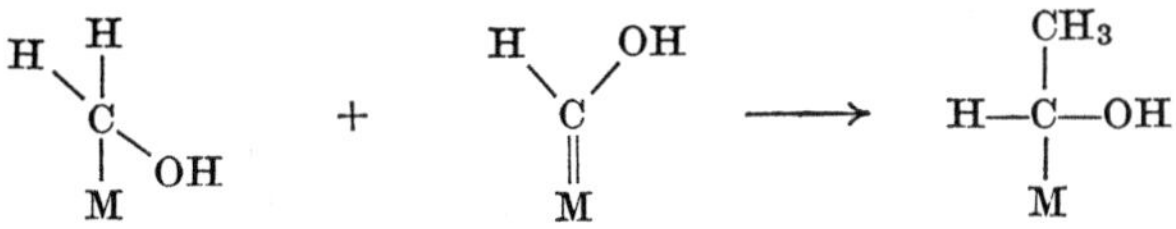

termination is most readily accomplished by hydrogenation.

Subject Index